SOTHIS PRESS

Sothis Press aims to completely reimagine the public domain book with an emphasis on universal access to captivatingly designed print editions. We believe that technology will enable every single book ever published to be perpetually accessible in beautiful print form. Sothis is beauty in ubiquity.

www.sothis.press
1 Waterhouse Square, London, ECIN 2ST

# The Elements of Non-Euclidean Geometry

Duncan M'laren
Young Sommerville

SOTHIS PRESS

ISBN 978-1-0365-1846-2



This edition is published by
Sothis Press © 2023. The company
is registered in England and Wales.
The Company Number is 14900389.

# THE ELEMENTS OF NON-EUCLIDEAN GEOMETRY

BY

D. M. Y. SOMMERVILLE, M.A., D.Sc.

CHICAGO
THE OPEN COURT PUBLISHING CO.
LONDON: G. BELL AND SONS, LTD.
1919

# PREFACE

THE present work is an extension and elaboration of a course of lectures on Non-Euclidean Geometry which I delivered at the Colloquium held under the auspices of the Edinburgh Mathematical Society in August, 1913.

Non-euclidean geometry is now a well-recognised branch of mathematics. It is the general type of geometry of homogeneous and continuous space, of which euclidean geometry is a special form. The creation or discovery of such types has destroyed the unique character of euclidean geometry and given it a setting amongst geometrical systems. There has arisen, so to speak, a science of Comparative Geometry.

Special care has, therefore, been taken throughout this book to show the bearing of non-euclidean upon euclidean geometry; and by exhibiting euclidean geometry as a really degenerate form—in the sense in which a pair of straight lines is a degenerate conic—to explain the apparent want of symmetry and the occasional failure of the principle of duality, which only a study of non-euclidean geometry can fully elucidate.

There are many ways of presenting the subject. In the present work the primary exposition follows the lines of elementary geometry in deduction from chosen postulates. This was the method of Euclid, and it was also the method of the discoverers of non-euclidean geometry. Restrictions have, however, been made. It was felt that a rigorously logical treatment, with a detailed examination of all the axioms or assumptions, would both overload the book and tend to render it dry and repulsive to the average reader. It is hoped, however, that the principles have been touched upon sufficiently to indicate the nature of the problems involved, especially in such cases where they throw light upon ordinary geometry.

It is impossible thoroughly to appreciate non-euclidean geometry without a knowledge of its history. I have therefore given in the first chapter a fairly full historical sketch of the subject up to the epoch of its discovery. Chapters II. and III. develop the principal results in hyperbolic and elliptic geometries. Chapter IV. gives the basis of an analytical treatment, the matter chosen for illustration here being, for the most part, such as was not touched on in the preceding chapters. This completes the rudiments of the subject. The next two chapters exhibit non-euclidean geometry in various lights, mathematical and philosophical, and bring up the history to a later stage. In the last three chapters some of the more interesting branches of geometry are worked out for the non-euclidean case, with a view to providing the serious student with a stimulus to pursue the subject in its higher developments. The reader will find a list of text-books and references to all the existing literature up to 1910 in my *Bibliography of Non-Euclidean Geometry* (London: Harrison, 1911).

Most of the chapters are furnished with exercises for working. As no examination papers in the subject are yet available, the examples have all been specially devised, or culled from original memoirs. Many of them are theorems of too special a character to be included in the text.

In preparing the treatise, the needs of the student reading privately have been kept steadily in view. Hence it is hoped that the work will prove useful to the "Scholarship Candidate" in our Secondary Schools who wishes to widen his geometrical horizon, to the Honours Student at our Universities who chooses Geometry as his special subject, and to the teacher of Geometry, in general, who desires to see in how far strict logical rigour can be made compatible with a treatment of the subject capable of comprehension by schoolboys.

In acknowledging my indebtedness to previous writers on the subject, special mention should be made of Bonola's article in the collection, *Questioni riguardanti la geometria elementare*, edited by Enriques (Bologna, 1900; German translation, Leipzig, 1911); and Liebmann's *Nicht-euklidische Geometrie* (Leipzig, 2nd ed., 1912).

I take this opportunity of expressing my obligations to Mr. Peter Fraser, M.A., B.Sc., Lecturer in Mathematics at the University of Bristol, and Mr. E. K. Wakeford, Trinity College, Cambridge, for kindly criticising the work while in manuscript form and giving many valuable suggestions.

I am also greatly indebted to Mr. W. P. Milne, M.A., D.Sc., Clifton College, Bristol, for continued assistance, by criticism and suggestion, all through the preparation of the book. To Dr. A. E. Taylor, Professor of Moral Philosophy at the University of St. Andrews, and Mr. C. D. Broad, B.A., Fellow of Trinity College, Cambridge, and Assistant to the Professor of Logic at the University of St. Andrews, I have also to express my thanks for reading and criticising Chapter VI. In correcting the proofs I have profited by the assistance of my wife and by the excellence of Messrs. MacLehose's printing work.

D. M. Y. S.

THE UNIVERSITY, ST. ANDREWS,
*April*, 1914.

# CONTENTS

## CHAPTER I.

### HISTORICAL.



## CHAPTER II.

### ELEMENTARY HYPERBOLIC GEOMETRY.

## CHAPTER III.

### ELLIPTIC GEOMETRY.

## CHAPTER IV.

### ANALYTICAL GEOMETRY.



## CHAPTER V.

### REPRESENTATIONS OF NON-EUCLIDEAN GEOMETRY IN EUCLIDEAN SPACE.

## CHAPTER VI.

### "SPACE-CURVATURE" AND THE PHILOSOPHICAL BEARING OF NON-EUCLIDEAN GEOMETRY.



## CHAPTER VII.

### RADICAL AXES, HOMOTHETIC CENTRES, AND SYSTEMS OF CIRCLES.

## CHAPTER VIII.

### INVERSION AND ALLIED TRANSFORMATIONS

## CHAPTER IX.

### THE CONIC.

# NON-EUCLIDEAN GEOMETRY

## CHAPTER I.

### HISTORICAL.

**1. The origins of geometry.**

Geometry, according to Herodotus, and the Greek derivation of the word, had its origin in Egypt in the mensuration of land, and the fixing of boundaries necessitated by the repeated inundations of the Nile. It consisted at first of isolated facts of observation and rude rules for calculation, until it came under the influence of Greek thought. The honour of having introduced the study of geometry from Egypt falls to THALES of Miletus (640-546 B.C.), one of the seven " wise men " of Greece. This marks the first step in the raising of geometry from its lowly level ; geometric elements were abstracted from their material clothing, and the geometry of lines emerged. With PYTHAGORAS (about 580-500 B.C.) geometry really began to be a metrical science, and in the hands of his followers and the succeeding Platonists the advance in geometrical knowledge was fairly rapid. Already, also, attempts were made, by HIPPOCRATES of Chios (about 430 B.C.) and others, to give a connected and logical presentation of the science in a series of propositions based upon a few axioms and definitions. The most famous of such attempts is, of course, that of EUCLID (about 300 B.C.), and so great was his prestige that he acquired, like

Aristotle, the reputation of infallibility, a fact which latterly became a distinct bar to progress.

## 2. Euclid's Elements.

The structure of Euclid's Elements should be familiar to every student of geometry, but owing to the multitude of texts and school editions, especially in recent years, when Euclid's order of the propositions has been freely departed from, Euclid's actual scheme is apt to be forgotten. We must turn to the standard text of Heiberg[1] in Greek and Latin, or its English equivalent by Sir Thomas Heath.[2]

Book I., which is the only one that immediately concerns us, opens with a list of definitions of the geometrical figures, followed by a number of postulates and common notions, called also by other Greek geometers "axioms."

Objection may be taken to many of the definitions, as they appeal simply to the intuition. The definition of a straight line as "a line which lies evenly with the points on itself" contains no statement from which we can deduce any propositions. We now recognise that we must start with some terms totally undefined, and rely upon postulates to assign a more definite character to the objects. A right angle and a square are defined before it has been shown that objects corresponding to the definitions can exist.

An axiom or common notion was considered by Euclid as a proposition which is so self-evident that it needs no demonstration; a postulate as a proposition which, though it may not be self-evident, cannot be proved by any simpler proposition. This distinction has been frequently misunderstood—to such an extent that later editors of Euclid have placed some of the postulates erroneously among the axioms. A notable instance is the parallel-postulate, No. 5, which has figured for ages as Axiom 11 or 12.

The common notions of Euclid are five in number, and deal exclusively with equalities and inequalities of magnitudes.

[1] 12 vols., Leipzig, 1883-99.  [2] 3 vols., Cambridge, 1908.

The postulates are also five in number and are exclusively geometrical. The first three refer to the construction of straight lines and circles. The fourth asserts the equality of all right angles, and the fifth is the famous Parallel-Postulate: "If a straight line falling on two straight lines make the interior angles on the same side less than two right angles, the two straight lines, if produced indefinitely, meet on that side on which are the angles less than two right angles."

### 3. Attempts to prove the parallel-postulate.

It seems impossible to suppose that Euclid ever imagined this to be self-evident, yet the history of the theory of parallels is full of reproaches against the lack of self-evidence of this "axiom." Sir Henry Savile[1] referred to it as one of the great blemishes in the beautiful body of geometry; D'Alembert[2] called it "l'écueil et le scandale des élémens de Géométrie."

The universal converse of the statement, "if two straight lines crossed by a transversal meet, they will make the interior angles on that side less than two right angles," is proved, with the help of another unexpressed assumption (that the straight line is of unlimited length), in Prop. 17; while the contrapositive, "if the interior angles on either side are not less than two right angles (*i.e.*, by Prop. 13, if they are equal to two right angles) the straight lines will not meet," is proved, again with the same assumption, in Prop. 28.

Such considerations induced geometers (and others), even up to the present day, to attempt its demonstration. From the invention of printing onwards a host of parallel-postulate demonstrators existed, rivalled only by the "circle-squarers," the "flat-earthers," and the candidates for the

[1] *Praelectiones*, Oxford, 1621 (p. 140).

[2] *Mélanges de Littérature*, Amsterdam, 1759 (p. 180).

Wolfskehl "Fermat" prize. Great ingenuity was expended, but no advance was made towards a settlement of the question, for each successive demonstrator showed the falseness of his predecessor's reasoning, or pointed out an unnoticed assumption equivalent to the postulate which it was desired to prove. Modern research has vindicated Euclid, and justified his decision in putting this great proposition among the independent assumptions which are necessary for the development of euclidean geometry as a logical system.

All this labour has not been fruitless, for it has led in modern times to a rigorous examination of the principles, not only of geometry, but of the whole of mathematics, and even logic itself, the basis of mathematics. It has had a marked effect upon philosophy, and has given us a freedom of thought which in former times would have received the award meted out to the most deadly heresies.

**4.** In a more restricted field the attempts of the postulate-demonstrators have given us an interesting and varied assortment of equivalents to Euclid's axiom. It would take up too much of our space to examine the numerous demonstrations,[1] but as some of the equivalent assumptions have come into school text-books, and there appears still to exist a belief that the Euclidean theory of parallels is a necessity of thought, it will be useful to notice a few of them.

One of the commonest of the equivalents used for Euclid's axiom in school text-books is "Playfair's axiom" (really due to Ludlam[2]): "Two intersecting straight lines cannot

[1] A useful account of these is given by W. B. Frankland in his *Theories of Parallelism*, Cambridge, 1910.

[2] *The Rudiments of Mathematics*, Cambridge, 1785 (p. 145).

both be parallel to the same straight line," which is equivalent to the statement, "Through a given point not more than one parallel can be drawn to a given straight line," and from this the properties of parallels follow very

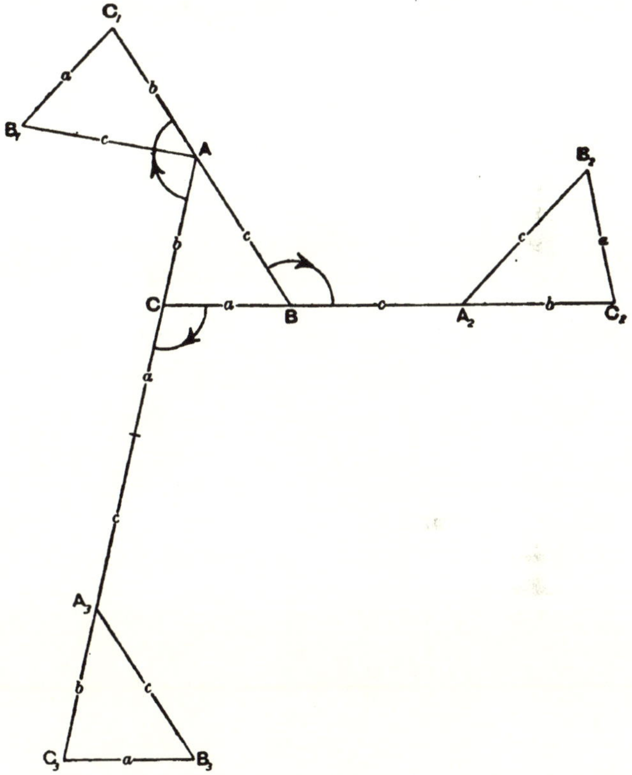


FIG. 1.

elegantly. The statement is simpler in form than Euclid's, but it is none the less an assumption.

Another equivalent is: "The sum of the angles of a triangle is equal to two right angles." I do not think that anyone has been so bold as to assume this as an axiom, but there have been many attempts to establish the theory of parallels by obtaining first an intuitive proof of this statement. A very neat proof, but particularly dangerous unless it be regarded merely as an illustration, is the "Rotation Proof," due to THIBAUT.[1]

5. Let a ruler (Fig. 1) be placed with its edge coinciding with a side $AC$ of a triangle, and let it be rotated successively about the three vertices $A$, $B$, $C$, in the direction $ABC$, so that it comes to coincide in turn with $AB$, $BC$ and $CA$. When it returns to its original position it must have rotated through four right angles. But this whole rotation is made up of three rotations through angles equal to the exterior angles of the triangle. The fault of this "proof" is that the three successive rotations are not equivalent at all to a single rotation through four right angles about a definite point, but are equivalent to a translation, through a distance equal to the perimeter of the triangle, along one of the sides.

The construction may be performed equally well on the surface of a sphere, with a ruler bent in the form of an arc of a great circle; and yet the sum of the exterior angles of a spherical triangle is always *less* than four right angles.

A similar fallacy is contained in all proofs based upon the idea of *direction*. Take the following: $AB$ and $CD$ (Fig. 2) are two parallel roads which are intersected by another road $BC$. A traveller goes along $AB$, and at $B$ turns into the road $BC$, altering his direction by the angle at $B$. At $C$ he turns into his original direction, and therefore must have turned back through the same angle. But this requires

[1] *Grundriss der reinen Mathematik*, 2nd ed. Göttingen, 1809.

a definition of sameness of direction, and this can only be effected when the theory of parallels has been established. The difficulty is made clear when we try to see what we mean by the relative compass-bearing of two points on the earth's surface. If we travelled due west from Plymouth along a parallel of latitude, we should arrive at Newfoundland, but the direct or shortest course would start in a "direction" WNW. and finish in the "direction" WSW.

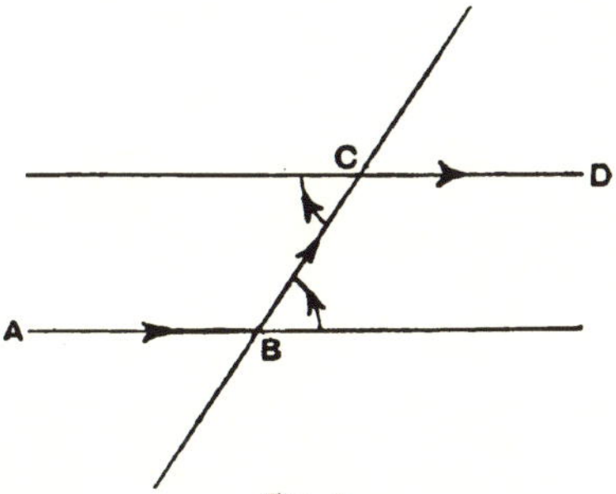


FIG. 2.

Other statements from which Euclid's postulate may be deduced are

"Three points are either collinear or concyclic." (W. Bolyai.[1])

"There is no upper limit to the area of a triangle." (Gauss.[2])

"Similar figures exist." (Wallis.[3])

**6**. Another class of demonstrations is based upon considerations of *infinite areas*. The following is "BERTRAND'S Proof."[4]

[1] *Kurzer Grundriss*, 1851 (p. 46).

[2] Letter to W. Bolyai, 16th December, 1799.

[3] *Opera*, Oxford, 1693 (t. ii. p. 676).

[4] L. Bertrand, *Développement nouveau de la partie élémentaire des mathématiques*, Geneva, 1778 (t. ii. p. 19).

Let a line $AX$ (Fig. 3), proceeding to infinity in the direction of $X$, be divided into equal parts $AB$, $BC$, ... and let the lines $AA'$, $BB'$, ... each produced to infinity, make equal angles with $AX$. Then the infinite strips $A'ABB'$, $B'BCC'$, ... can all be superposed and have equal areas, but it requires infinitely many of these strips to make up the area $A'AX$, contained between the lines $AA'$ and $AX$, each produced to infinity. Again, let the angle $A'AX$ be divided into equal parts $A'AP$, $PAQ$, .... Then all these sectors can be superposed and have equal areas, but it requires only a finite number of them to make up the area $A'AX$.

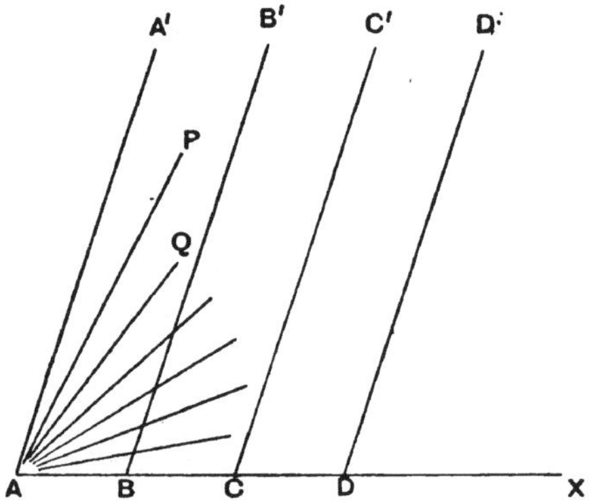


FIG. 3.

Hence, however small the angle $A'AP$ may be, the area $A'AP$ is greater than the area $A'ABB'$, and cannot therefore be contained within it. $AP$ must therefore cut $BB'$; and this result is easily recognised as Euclid's axiom.

The fallacy here consists in applying the principle of superposition to infinite areas, as if they were finite magnitudes.

If we consider (Fig. 4) two infinite rectangular strips $A'ABB'$ and $A'PQB'$ with equal bases $AB$, $PQ$, and partially superposed, then the two strips are manifestly unequal, or else the principle of superposition is at fault. Again, suppose we have two rectangular strips $A'ABB'$, $C'CDD'$ (Fig. 5). Mark off equal lengths $AA_1$, $A_1A_2$, ... along $AA'$, each equal to $CD$, and equal lengths $CC_1$, $C_1C_2$, ... along $CC'$, each equal to $AB$, and divide the strips at these points into rectangles. Then all the rectangles are equal, and, if we

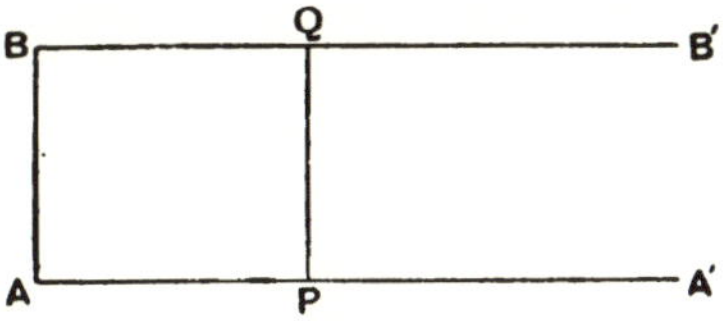


FIG. 4.

number them consecutively, then to every rectangle in the one strip there corresponds the similarly numbered rectangle in the other strip. Hence, if the ordinary theorems of congruence and equality of areas are assumed, we must admit that the two strips are equal in area, and that therefore the area is independent of the magnitude of $AB$. Such deductions are just as valid as the deduction of Euclid's axiom from a consideration of infinite areas.

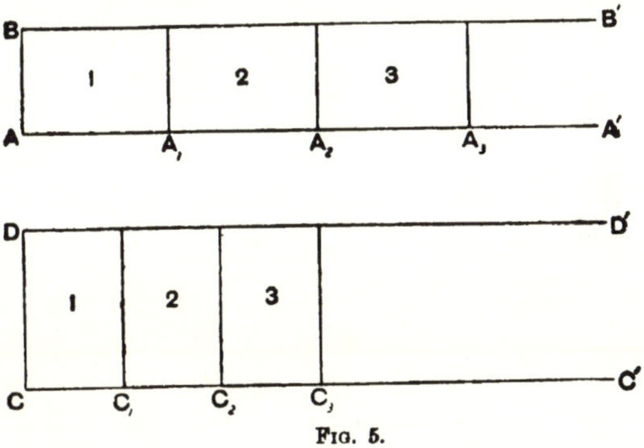


FIG. 5.

**7.** It will suffice to give one other example of the attempts to base the theory of parallels on intuition. Suppose that, instead of Euclid's definition of parallels as "straight lines, which, being in the same plane, and being produced indefinitely in both directions, do not meet one another in either direction," we define them as "straight lines which are everywhere equidistant," then the whole Euclidean theory of parallels comes out with beautiful simplicity. In particular, the sum of the angles of any triangle $ABC$ (Fig. 6) is proved equal to two right angles by drawing through the vertex $A$ a parallel to the base $BC$. Then, if we draw perpendiculars from $A$, $B$, $C$ on the opposite parallel, these perpendiculars are all equal. The angle $EAB = \angle B$ and the angle $CAF = \angle C$.

It is scarcely necessary to point out, however, that this definition contains the whole debatable assumption. We

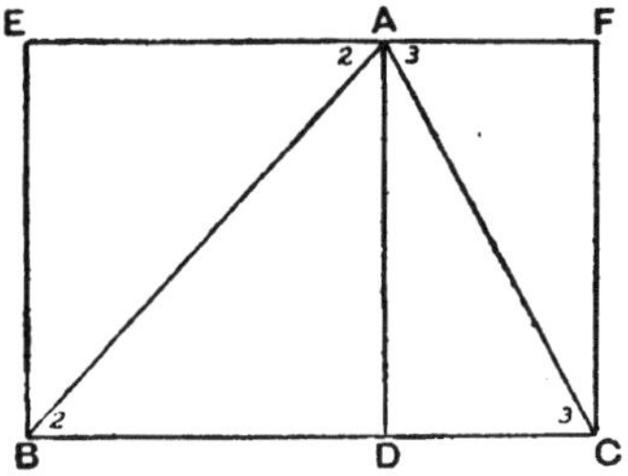


FIG. 6.

have no warrant for assuming that a pair of straight lines can exist with the property ascribed to them in the definition. To put it another way, if a perpendicular of constant length move with one extremity on a fixed line, is the locus of its free extremity another straight line? We shall find reason later on to doubt this. In fact, non-euclidean geometry has made it clear that the ideas of parallelism and equidistance are quite distinct. The term "parallel" Greek παράλληλος = running alongside) originally connoted equidistance, but the term is used by Euclid rather in the sense "asymptotic" (Greek ἀ-σύμπτωτος = non-intersecting), and this term has come to be used in the limiting case of curves which tend to coincidence, or the limiting case between intersection and non-intersection. In non-euclidean geometry parallel straight lines are asymptotic in this sense, and equidistant straight lines in a plane do not exist. This is just one instance of two distinct ideas which are confused in euclidean geometry, but are quite distinct in non-euclidean. Other instances will present themselves.

**8. First glimpses of Non-Euclidean geometry.**

Among the early postulate-demonstrators there stands a unique figure, that of a Jesuit, Gerolamo SACCHERI (1667-1733), contemporary and friend of Ceva. This man devised an entirely different mode of attacking the problem, in an attempt to institute a *reductio ad absurdum*.[1] At that time the favourite starting-point was the conception of parallels as equidistant straight lines, but Saccheri, like some of his predecessors, saw that it would not do to assume this in the definition. He starts with two equal perpendiculars $AC$ and $BD$ to a line $AB$. When the ends $C$, $D$ are joined, it is easily proved that the angles at $C$ and $D$ are equal; but are they right angles? Saccheri keeps an open mind, and proposes three hypotheses:

(1) The Hypothesis of the Right Angle.
(2) The Hypothesis of the Obtuse Angle.
(3) The Hypothesis of the Acute Angle.

The object of his work is to demolish the last two hypotheses and leave the first, the Euclidean hypothesis, supreme;

[1] *Euclides ab omni naevo vindicatus*, Milan, 1733. English trans. by Halsted, *Amer. Math. Monthly*, vols. 1-5, 1894-98; German by Stäckel and Engel, *Die Theorie der Parallellinien*, Leipzig, 1895. (This book by Stäckel and Engel contains a valuable history of the theory of parallels.)

but the task turns out to be more arduous than he expected. He establishes a number of theorems, of which the most important are the following:

If one of the three hypotheses is true in any one case, the same hypothesis is true in every case.

On the hypothesis of the right angle, the obtuse angle, or the acute angle, the sum of the angles of a triangle is equal to, greater than, or less than two right angles.

On the hypothesis of the right angle two straight lines intersect, except in the one case in which a transversal cuts them at equal angles. On the hypothesis of the obtuse angle two straight lines always intersect. On the hypothesis of the acute angle there is a whole pencil of lines through a given point which do not intersect a given straight line, but have a common perpendicular with it, and these are separated from the pencil of lines which cut the given line by two lines which approach the given line more and more closely, and meet it at infinity.

The locus of the extremity of a perpendicular of constant length which moves with its other end on a fixed line is a straight line on the first hypothesis, but on the other hypotheses it is curved; on the hypothesis of the obtuse angle it is convex to the fixed line, and on the hypothesis of the acute angle it is concave.

Saccheri demolishes the hypothesis of the obtuse angle in his Theorem 14 by showing that it contradicts Euclid I. 17 (that the sum of any two angles of a triangle is less than two right angles); but he requires nearly twenty more theorems before he can demolish the hypothesis of the acute angle, which he does by showing that two lines which meet in a point at infinity can be perpendicular at that point to the same straight line. In spite of all his efforts, however, he does not seem to be quite satisfied with the validity of his proof, and he offers another proof in which he loses himself, like many another, in the quicksands of the infinitesimal.

If Saccheri had had a little more imagination and been less bound down by tradition, and a firmly implanted belief that Euclid's hypothesis was the only true one, he would have anticipated by a century the discovery of the two non-euclidean geometries which follow from his hypotheses of the obtuse and the acute angle.

9. Another investigator, J. H. LAMBERT (1728-1777),[1] fifty years after Saccheri, also fell just short of this discovery. His starting-point is very similar to Saccheri's, and he distinguishes the same three hypotheses; but he went further than Saccheri. He actually showed that on the hypothesis of the obtuse angle the area of a triangle is proportional to the excess of the sum of its angles over two right angles, which is the case for the geometry on the sphere, and he concluded that the hypothesis of the acute angle would be verified on a sphere of imaginary radius. He also made the noteworthy remark that on the third hypothesis there is an absolute unit of length which would obviate the necessity of preserving a standard foot in the Archives.

He dismisses the hypothesis of the obtuse angle, since it requires that two straight lines should enclose a space, but his argument against the hypothesis of the acute angle, such as the non-existence of similar figures, he characterises

[1] *Theorie der Parallellinien*, 1786. (Reprinted in Stäckel and Engel, *Th. der Par.*, 1895.)

as arguments *ab amore et invidia ducta.* Thus he arrived at no definite conclusion, and his researches were only published some years after his death.

**10.** About this time (1799) the genius of GAUSS (1777-1855) was being attracted to the question, and, although he published nothing on the subject except a few reviews, it is clear from his correspondence and fragments of his notes that he was deeply interested in it. He was a keen critic of the attempts made by his contemporaries to establish the theory of parallels; and while at first he inclined to the orthodox belief, encouraged by Kant, that Euclidean geometry was an example of a necessary truth, he gradually came to see that it was impossible to demonstrate it. He declares that he refrained from publishing anything because he feared the clamour of the Boeotians, or, as we should say, the Wise Men of Gotham; indeed at this time the problem of parallel lines was greatly discredited, and anyone who occupied himself with it was liable to be considered as a crank.

Gauss was probably the first to obtain a clear idea of the possibility of a geometry other than that of Euclid, and we owe the very name Non-Euclidean Geometry to him.[1] It is clear that about the year 1820 he was in possession of many theorems of non-euclidean geometry, and though he meditated publishing his researches when he had sufficient leisure to work them out in detail with his characteristic elegance, he was finally forestalled by receiving in 1832, from his friend W. Bolyai, a copy of the now famous Appendix by his son, John Bolyai.

**11**. Among the contemporaries and pupils of Gauss there are three names which deserve mention. F. K. SCHWEIKART (1780-1859),

[1] Letter to Taurinus, 8th November, 1824.

Professor of Law in Marburg, sent to Gauss in 1818 a page of MS. explaining a system of geometry which he calls "Astral Geometry," in which the sum of the angles of a triangle is always less than two right angles, and in which there is an absolute unit of length.

He did not publish any account of his researches, but he induced his nephew, F. A. TAURINUS (1794-1874), to take up the question. His uncle's ideas did not appeal to him, however, but a few years later he attempted a treatment of the theory of parallels, and having received some encouragement from Gauss, he published a small book, *Theorie der Parallellinien*, in 1825. After its publication he came across Camerer's new edition of Euclid in Greek and Latin, which, in an Excursus to Euclid I. 29, contains a very valuable history of the theory of parallels, and there he found that his methods had been anticipated by Saccheri and Lambert. Next year, accordingly, he published another work, *Geometriae prima elementa*, and in the Appendix to this he works out some of the most important trigonometrical formulae for non-euclidean geometry by using the fundamental formulae of spherical geometry with an imaginary radius. Instead of the notation of hyperbolic functions, which was then scarcely in use, he expresses his results in terms of logarithms and exponentials, and calls his geometry the "Logarithmic Spherical Geometry."

Though Taurinus must be regarded as an independent discoverer of non-euclidean trigonometry, he always retained the belief, unlike Gauss and Schweikart, that Euclidean geometry was necessarily the true one. Taurinus himself was aware, however, of the importance of his contribution to the theory of parallels, and it was a bitter disappointment to him when he found that his work attracted no attention. In disgust he burned the remainder of the edition of his *Elementa*, which is now one of the rarest of books.

The third to be mentioned as having arrived at the notion of a geometry in which Euclid's postulate is denied is F. L. WACHTER (1792-1817), a student under Gauss. It is remarkable that he affirms that even if the postulate be denied, the geometry on a sphere becomes identical with the geometry of Euclid when the radius is indefinitely increased, though it is distinctly shown that the limiting surface is not a plane. This was one of the greatest discoveries of Lobachevsky and Bolyai. If Wachter had lived he might have been the discoverer of non-euclidean geometry, for his insight into the question was far beyond that of the ordinary parallel-postulate demonstrator.

**12.** While Gauss, Schweikart, Taurinus and others were working in Germany, and had arrived independently at some of the results of non-euclidean geometry, and were, in fact, just on the threshold of its discovery, in France and Britain the ideas were still at the old stage, though there was a considerable interest in the subject, inspired chiefly by A. M. Legendre (1752-1833). Legendre's researches were published in the various editions of his *Éléments*, from 1794 to 1823, and collected in an extensive article in the Memoirs of the Paris Academy in 1833.

Assuming all Euclid's definitions, axioms and postulates, except the parallel-postulate and all that follows from it, he proves some important theorems, two of which, Propositions A and B, are frequently referred to in later work as Legendre's First and Second Theorems.

Prop. A. *The sum of the three angles of a rectilinear triangle cannot be greater than two right angles* ($\pi$). (*Éléments*, 3rd ed. 1800.)

In Fig. 7, $A_0A_1A_2 \dots A_n$ is a straight line, and the triangles $A_0B_0A_1$, $A_1B_1A_2$, ... are all congruent, and the vertices $B_0B_1 \dots B_n$ are joined by a broken line.

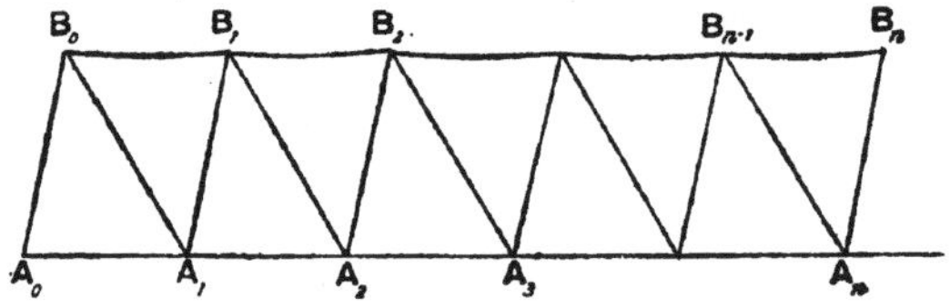


Fig. 7.

Suppose, if possible, that

$$\angle A_0B_0A_1 + B_0A_0A_1 + A_0A_1B_0 > \pi.$$

Now $\angle B_0A_0A_1 = B_1A_1A_2$

and $$\angle B_0A_1B_1 + B_1A_1A_2 + A_0A_1B_0 = \pi.$$

Therefore $\angle A_0B_0A_1 > B_0A_1B_1$,
and therefore $A_0A_1 > B_0B_1$.

Let $A_0A_1 - B_0B_1 = d$; then

$$A_0B_0 + B_0B_1 + B_1B_2 + \ldots + B_{n-1}B_n + B_nA_n = 2A_0B_0 + nB_0B_1$$
$$= 2A_0B_0 + nA_0A_1 - nd = A_0A_n + 2A_0B_0 - nd,$$

*i.e.* $A_0A_n = A_0B_0 + B_0B_1 + \ldots + B_nA_n + (nd - 2A_0B_0)$.

But, by increasing $n$, $nd$ can be made to exceed the fixed length $2A_0B_0$; and hence $A_0A_n$, which is the length of the straight line joining $A_0$ and $A_n$, can be greater than the sum of the parts of the broken line which joins the same two points, which is absurd.

There are several points in this proof that require careful examination.

In the first place, the assumption that $nd$ can always exceed $2A_0B_0$ by taking $n$ sufficiently great lies at the basis of geometrical *continuity*, and is equivalent to the denial of the existence of infinitesimals. This is generally known as the *Axiom of Archimedes.* The question of continuity is fundamental in dealing with the foundations of geometry, but it would be outside the scope of this book to enter further into this extensive and difficult subject.

Twice in this proof we have assumed the "theorem of the exterior angle" of a triangle (Euclid I. 16), first in the statement that $A_0A_1 > B_0B_1$, and second in the assumption that the straight line joining two points is the shortest path (Euclid I. 20). This is equivalent to the rejection of Saccheri's hypothesis of the obtuse angle. If this hypothesis be followed to its logical conclusion, it can be shown (see Chap. III.) that two straight lines in a plane will always intersect, when produced in *either* direction. The straight line is then *re-entrant*, and there are at least two straight paths connecting any two points. The straight line $A_0A_n$ would not then of necessity be the shortest path from $A_0$ to $A_n$.

PROP. B. *If there exists a single triangle in which the sum of the angles is equal to two right angles, then in every triangle the sum of the angles must likewise be equal to two right angles.*

This proposition was already proved by Saccheri, along with the corresponding theorem for the case in which the sum of the angles is less than two right angles, and we need not reproduce Legendre's proof, which proceeds by constructing successively larger and larger triangles, in each of which the sum of the angles $= \pi$.

Legendre makes an attempt to prove that *the sum of the angles of a triangle is equal to two right angles*, as follows (*Éléments*, 12th ed. 1823):

Let $A_1B_1C_1$ (Fig. 8) be a triangle, in which $A_1B_1$ is the greatest side and $B_1C_1$ the least. Join $A_1$ to $M_1$, the middle

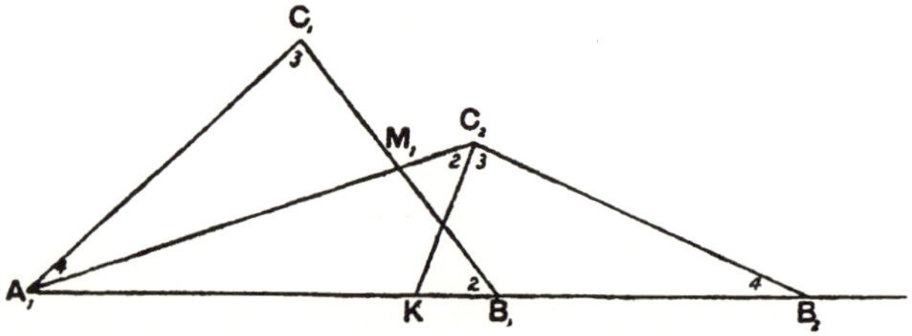


FIG. 8.

point of $B_1C_1$, and produce $A_1M_1$ to $C_2$ so that $A_1C_2 = A_1B_1$. On $A_1B_1$ take $A_1K = KB_2 = A_1M_1$, and join $C_2K$. Then we get a second triangle $A_2B_2C_2$, in which $A_2$ coincides with $A_1$, and in which $A_2B_2$ is the greatest side and $B_2C_2$ the least. Denote the angles of the triangles $A_1B_1C_1$, $A_2B_2C_2$ by single letters.

Then $\triangle A_1C_2K \equiv A_1B_1M_1$ and $\triangle C_2KB_2 \equiv C_1M_1A_1$.

Therefore $\angle A_1C_2K = B_1, \ \angle KC_2B_2 = C_1, \ \angle M_1A_1C_1 = B_2$.

Therefore $A_1 = A_2 + B_2$ and $B_1 + C_1 = C_2$.

Hence $A_1+B_1+C_1=A_2+B_2+C_2$

and area $A_1B_1C_1=$ area $A_2B_2C_2$.

By repeating this construction we get a series of triangles with the same area and angle-sum.

Now $A_2<\frac{1}{2}A_1$, $A_3<\frac{1}{2}A_2<\frac{1}{4}A_1, \ldots,$ $A_{n+1}<\frac{1}{2^n}A_1$,

$$B_2<A_1,\quad B_3<A_2<\tfrac{1}{2}A_1, \ldots,\quad B_{n+1}<\frac{1}{2^{n-1}}A_1.$$

Hence the angles $A_n$ and $B_n$ both tend to zero, while the vertex $C_n$ ultimately lies on $A_nB_n$. The sum of the angles thus reduces to the single angle $C_n$, which is ultimately equal to two right angles.

In this proof there is a latent assumption and also a fallacy. In the first place it is tacitly assumed that the straight line is not re-entrant, for if it were re-entrant the "theorem of the exterior angle," upon which the proofs of the inequalities depend, could not be accepted, and the whole proof is invalidated. Again, if we grant the theorem of the exterior angle, $B_n$ and $C_n$ both go to infinity, and we cannot draw any conclusions as regards the magnitude of the angle $C_n$.

Legendre's other attempts make use of infinite areas. He makes reference to Bertrand's proof, and attempts to prove the necessity of Playfair's axiom in this way: if it be denied, then a straight line would be contained entirely within the angle formed by two rays, but this is impossible since the area enclosed by the angle is less than "half the area of the whole plane."

13. In Britain the investigations of Legendre stimulated such men as Playfair and Leslie (Professors at Edinburgh), Ivory, Perronet Thompson, and Henry Meikle. Of these, however, none but Meikle had advanced beyond

the stage of Legendre. Meikle [1] actually proved in detail, what had been foreshadowed fifty years before by Lambert, that if the sum of the angles of a triangle is less than two right angles the defect is proportional to its area. He rejected the hypothesis because he would not admit the existence of a triangle with all its angles zero. He also proved independently Saccheri's general form of Legendre's second theorem.

But by this time the epoch-making works of Lobachevsky and Bolyai had been published, and the discovery of a logically consistent system of geometry in which the parallel-postulate is denied proved once for all that all attempts to deduce this postulate from the other axioms are doomed to failure. It was not, however, in Germany after all that non-euclidean geometry at last saw the light, but simultaneously in remote districts of Russia and Hungary.

### 14. The discovery of Non-Euclidean geometry.

Nikolai Ivanovich LOBACHEVSKY (1793-1856), Professor of Mathematics at Kazan, was interested in the theory of parallels from at least 1815. Lecture notes of the period 1815-17 are extant, in which Lobachevsky attempts in various ways to establish the Euclidean theory. He proves Legendre's two propositions, and employs also the ideas of direction and infinite areas. In 1823 he prepared a treatise on geometry for use in the University, but it obtained so unfavourable a report that it was not printed. The MS. remained buried in the University Archives until it was discovered and printed in 1909. In this book he states that " a rigorous proof of the postulate of Euclid has

[1] *Edinburgh New Philos. Journ.*, **36** (1844), p. 313.

not hitherto been discovered; those which have been given may be called explanations, and do not deserve to be considered as mathematical proofs in the full sense."

Just three years afterwards, he read to the physical and mathematical section of the University of Kazan a paper entitled "Exposition succinte des principes de la géométrie avec une démonstration rigoureuse du théorème des parallèles." In this paper, the manuscript of which has unfortunately been lost, Lobachevsky explains the principles of his "Imaginary Geometry," which is more general than Euclid's, and in which two parallels can be drawn to a given line through a given point, and in which the sum of the angles of a triangle is always less than two right angles.

In the course of a busy life Lobachevsky wrote some half dozen extensive memoirs expounding the new geometry. The first of these were in Russian, and therefore inaccessible. In 1840 he tried to reach a wider circle with a small book in German entitled *Geometrische Untersuchungen zur Theorie der Parallellinien*, and just before his death he wrote a summary of his researches under the title "Pangeometry," which he put into French and contributed to the memorial volume published at the jubilee of his own University.[1]

**15.** Bolyai János (John) (1802-1860) was the son of Bolyai Farkas (Wolfgang) (1775-1856), a fellow-student and friend of Gauss at Göttingen. The father was early interested in the theory of parallels, and without doubt discussed the subject with Gauss while at Göttingen. The professor of mathematics at that time, A. G. Kaestner, had

[1] An English translation of the *Geometrische Untersuchungen* was published by Halsted (Austin, Texas, 1891). An extensive Life of Lobachevsky was published, together with German translations of two of the Russian papers, by Engel (Leipzig, 1898).

himself attacked the problem, and with hi help G. S. Klügel, one of his pupils, compiled in 1763 the earliest history of the theory of parallels.

In 1804 Wolfgang Bolyai, just after his appointment as professor of mathematics in Maros-Vásárhely, sent to Gauss a "Theory of Parallels," the elaboration of his Göttingen studies. In this he gives a demonstration very similar to that of Meikle and some of Perronet Thompson's, in which he tries to prove that a series of equal segments placed end to end at equal angles, like the sides of a regular polygon, must make a complete circuit. Though Gauss clearly revealed the fallacy, Bolyai persevered and sent Gauss, in 1808, a further elaboration of his proof. To this Gauss did not reply, and Bolyai, wearied with his ineffectual endeavours to solve the riddle of parallel lines, took refuge in poetry and composed dramas. During the next twenty years, amid various interruptions, he put together his system of mathematics, and at length, in 1832-3, published in two volumes an elementary treatise[1] on mathematical discipline which contains all his ideas with regard to the first principles of geometry.

Meanwhile John Bolyai, while a student at the Royal College for Engineers at Vienna, had been giving serious attention to the theory of parallels, in spite of his father's solemn adjuration to let the loathsome subject alone. At first, like his predecessors, he attempted to find a proof for the parallel-postulate, but gradually, as he focussed his attention more and more upon the results which would follow from a denial of the axiom, there developed in his mind the idea of a general or "Absolute Geometry" which

[1] *Tentamen juventutem studiosam in elementa matheseos . . . introducendi*, Maros-Vásárhely, 1832-3.

would contain ordinary or euclidean geometry as a special or limiting case. Already, in 1823, he had worked out the main ideas of the non-euclidean geometry, and in a letter of 3rd November he announces to his father his intention of publishing a work on the theory of parallels, "for," he says, "I have made such wonderful discoveries that I am myself lost in astonishment, and it would be an irreparable loss if they remained unknown. When you read them, dear Father, you too will acknowledge it. I cannot say more now except that out of nothing I have created a new and another world. All that I have sent you hitherto is as a house of cards compared to a tower." Wolfgang advised his son, if his researches had really reached the desired goal, to get them published as soon as possible, for new ideas are apt to leak out, and further, it often happens that a new discovery springs up spontaneously in many places at once, "like the violets in springtime." Bolyai's presentment was truer than he suspected, for at this very moment Lobachevsky at Kazan, Gauss at Göttingen, Taurinus at Cologne, were all on the verge of this great discovery It was not, however, till 1832 that at length the work was published. It appeared in Vol. I. of his father's *Tentamen*, under the title "Appendix, scientiam absolute veram exhibens."

W. Bolyai wrote one other book,[1] in German, in which he refers to the subject, but the son, although he continued to work at his theory of space, published nothing further. Lobachevsky's *Geometrische Untersuchungen* came to his

[1] *Kurzer Grundriss eines Versuchs*, Maros-Vásárhely, 1851. J. Bolyai's "Appendix" has been translated into French, Italian, German, English and Magyar; English by Halsted (Austin, Texas, 1891). A complete life of the Bolyai, with German translations of the "Appendix," parts of the *Tentamen*, etc., has been published by Stäckel (Leipzig, 1913), as a companion book to Engel's *Lobatschefskij*.

knowledge in 1848, and this spurred him on to complete the great work on "Raumlehre," which he had already planned at the time of the publication of his "Appendix," but he left this in large part as a *rudis indigestaque moles*, and he never realised his hope of triumphing over his great Russian rival.

On the other hand, Lobachevsky never seems to have heard of Bolyai, though both were directly or indirectly in communication with Gauss. Much has been written on the relationship of these three discoverers, but it is now generally recognised that John Bolyai and Lobachevsky each arrived at their ideas independently of Gauss and of each other; and, since they possessed the convictions and the courage to publish them which Gauss lacked, to them alone is due the honour of the discovery.

**16.** The succeeding history of non-euclidean geometry will be passed over here very briefly.[1] The ideas inaugurated by Lobachevsky and Bolyai did not for many years attain any wide recognition, and it was only after Baltzer had called attention to them in 1867, and at his request Hoüel had published French translations of the epoch-making works, that the subject of non-euclidean geometry began to be seriously studied.

It is remarkable that while Saccheri and Lambert both considered the two hypotheses, it never occurred to Lobachevsky or Bolyai or their predecessors, Gauss,

[1] Some of the later history will be given in Chap. VI. The best history of the subject is R. Bonola: *La geometria non-euclidea: esposizione storico-critica del suo sviluppo* (Bologna, 1906); English translation (based on the German translation by Liebmann, Leipzig, 1908) by H. S. Carslaw (Chicago, 1912). A full classified bibliography is to be found in Sommerville's *Bibliography of non-euclidean geometry, including the theory of parallels, the foundations of geometry and space of n dimensions* (London, 1911).

Schweikart, Taurinus and Wachter, to admit the hypothesis that the sum of the angles of a triangle may be greater than two right angles. This involves the conception of a straight line as being unbounded but yet of finite length. Somewhere " at the back of beyond " the two ends of the line meet and close it. We owe this conception first to Bernhard RIEMANN (1826-1866) in his Dissertation of 1854 [1] (published only in 1866 after the author's death), but in his Spherical Geometry two straight lines intersect twice like two great circles on a sphere. The conception of a geometry in which the straight line is finite, and is, without exception, uniquely determined by two distinct points, is due to Felix KLEIN.[2] Klein attached the now usual nomenclature to the three geometries ; the geometry of Lobachevsky he called *Hyperbolic*, that of Riemann *Elliptic*, and that of Euclid *Parabolic*.

## EXAMPLES I.

**1.** If the angle in a semicircle is constant, prove that it is a right angle.

**2.** $AB$ is a fixed line and $P$ a variable point such that the angle $APB$ is constant. Show that the tangents at $A$ and $B$ to the curve locus of $P$ are equally inclined to $AB$.

**3.** If every chord in the locus of Question 2 has the property that it subtends a constant angle at points on the curve, prove that the sum of the angles of a triangle must be equal to two right angles.

Examine the fallacies in the following proofs of Euclid's axiom :

**4.** If the side $c$ and the angles $A$ and $B$ of a triangle are given the triangle is determined, and therefore the angle $C=f(A, B, c)$. But since this equation must be homogeneous, it cannot contain the

[1] " Über die Hypothesen, welche der Geometrie zu Grunde liegen " ; English translation by W. K. Clifford, *Nature*, **8** (1873).

[2] " Über die sogenannte Nicht-Euklidische Geometrie," *Math. Annalen*, **4** (1871), **6** (1873).

side $c$. Hence $C=f(A, B)$. Let $ABC$ be a right-angled triangle, and draw the perpendicular $CD$ on the hypotenuse. Then the two triangles $ABC$, $ACD$ have two angles equal, and therefore the third angle $ACD=B$. Similarly $BCD=A$. Therefore $A+B+C$ $=2$ right angles. (A. M. Legendre, 1794.)

**5.** Let $OB$ and $NA$ be both perpendicular to $ON$, then $OB$ does not meet $NA$. Let $OG$, making a finite angle $GOB$, be the last line through $O$ which meets $NA$. Then $NA$ can be produced beyond its point of intersection with $OG$ to $K$, and $OK$ still meets $NA$. Hence $OG$ is not the last, and therefore all lines through $O$ within the angle $NOB$ must meet $NA$. (J. D. Gergonne, 1812.)

**6.** One altitude $AD$ of an equilateral triangle $ABC$ divides it into two right-angled triangles, in which one acute angle is double the other. If the three altitudes meet in $O$, each of the triangles $AOE$, etc., has one angle equal to half the angle of the equilateral triangle; hence the angle $OAE=\frac{1}{2}AOE$. Hence the sum of the angles of the triangle $ABC$ is equal to half the sum of the angles at $O$, *i.e.* equal to two right angles. (J. K. F. Hauff, 1819.)

**7.** $AA'\perp AB$ and $ABB'$ is acute. From $D$, any point on $BB'$, is drawn the perpendicular $DE$ to $AB$. $C$ is any point on $AA'$, and $BC$ cuts $DE$ in $F$. $G$ is the middle point of $EF$, and $BG$ meets $AC$ in $H$. An isosceles triangle is drawn with base $EF$ and sides equal to $ED$, making the base angles $=\alpha$. Rotate the plane of the figure about $AB$ through the angle $\alpha$. Denote the points in their new positions by suffixes. Then $D_1G\perp EF$, and $BH$ is the projection of $BB_1'$. $H$ is therefore the projection of a point on both $BB_1'$ and $AA_1'$, and these lines therefore meet. (K. Th. E. Gronau, 1902.)

# CHAPTER II.

## ELEMENTARY HYPERBOLIC GEOMETRY.

### 1. Fundamental assumptions.

In establishing any system of geometry we must start by naming certain objects which we cannot define in terms of anything more elementary, and make certain assumptions, from which by the laws of logic we can develop a consistent system. These assumptions are the axioms of the science. The axioms of geometry have been classified by Hilbert [1] under five groups :

1. *Axioms of connection*, or classification, connecting point, line and plane.
2. *Axioms of order*, explaining the idea of "between."
3. *Axioms of congruence.*
4. *Axiom of parallels.*
5. *Axioms of continuity.*

Without entering into these in more detail,[2] we shall assume, as deductions from them, the theorems relating to the comparison and addition of segments and angles.

The method of superposition can be used as a *façon de parler*. Strictly speaking, a geometrical figure is incapable of being *moved* ; [3]

[1] D. Hilbert, *Grundlagen der Geometrie*, Leipzig, 1899, 4th ed. 1913 ; English translation by Townsend, Chicago, 1902.

[2] The reader who wishes to study the development of non-euclidean geometry from a set of axioms may refer to J. L. Coolidge, *Elements of Non-Euclidean Geometry*, Oxford, 1909.

[3] Cf. Chap. VI. § 4.

lines are not *drawn*, nor are figures *constructed*. It is only the act of the mind which fixes the attention on certain geometrical figures which already exist, developing them out, so to speak, like the picture on a photographic plate. And when we speak of applying one figure to another by superposition, all that we mean is that a comparison is made between the two figures and certain results deduced by the axioms of congruence. When a geometrical figure, *e.g.* a line or a point, is spoken of as moving, we are really transferring our attention to a succession of lines or points in different positions.

The measurement of angles is independent of the theory of parallels. Vertically opposite angles are equal; the sum of the four angles made by two intersecting lines is an absolute constant, and one quarter of this is a right angle. An absolute unit of angle, therefore, exists. A "flat-angle," which is equal to two right angles, is generally denoted by the symbol $\pi$. Through a given point only one perpendicular can be drawn to a given straight line, the usual construction for this being always possible.

The question of the numerical value of $\pi$, or, what is the same thing, of the unit of angle, need not concern us until we come to consider trigonometrical formulae (see § 39). We may, however, state at once that when $\pi$ is treated as a number it has just the value which we are already accustomed to assign to it, viz. $3\frac{1}{7}$ to a rough approximation, or, accurately, 4 times the sum of the infinite series $1 - \frac{1}{3} + \frac{1}{5} - \frac{1}{7} + \ldots$. But it is necessary to warn the reader that $\pi$ does not stand for the ratio of the circumference of a circle to its diameter, for in non-euclidean geometry this ratio is not constant; and the radian, or unit angle, in terms of which a flat-angle is represented by the number $\pi$, does not admit of the familiar construction by means of a circle.

We shall assume, as deductions from the axioms of congruence, the congruence-theorems for triangles (Euc. I. 4, 8, 26), and those on the base-angles of an isosceles triangle (Euc. I. 5, 6), which imply the symmetry of the plane. The theorems relating to inequalities among the sides and angles of a triangle (Euc. I. 16-20) are true within a restricted region. In particular, the "theorem of the exterior angle," upon which the others depend, is proved in § 6 to be true without exception in hyperbolic geometry.

An important axiom of order which must be explicitly mentioned is PASCH'S AXIOM.[1] *If a straight line cuts one side of a triangle and does not pass through a vertex, it will also cut one of the other sides* ("side" being understood to mean the segment subtended by the opposite interior angle of the triangle).

A large part of geometry can be constructed without the axioms of continuity,[2] but we shall in general assume continuity.

The watershed, so to speak, between the euclidean and the non-euclidean geometries which we are about to develop, is the axiom of parallels.

## 2. Parallel lines.

Consider (Fig. 9) a straight line $l$ and a point $O$ not on the line. Let $ON$ be $\perp$ $l$, and take any point $P$ on $l$. The line $OP$ cuts $l$ in $P$. As the point $P$ moves along $l$ away from $N$ there are two possibilities to consider:

(1) $P$ may return to its starting point after having traversed a finite distance. This is the hypothesis of ELLIPTIC GEOMETRY.

(2) $P$ may continue moving, and the distance $NP$ tend to infinity. This hypothesis is true in ordinary geometry. The ray $OP$ then tends to a definite limiting[3] position $OL$,

[1] M. Pasch, *Vorlesungen über neuere Geometrie*, Leipzig, 1882; 2nd ed. 1912.

[2] Cf. G. B. Halsted, *Rational Geometry*, New York, 1904.

[3] This assumes continuity. We might dispense with this assumption by assuming a definite line $OL$ which separates the intersectors of $NA$ from the non-intersectors.

and $OL$ is said to be *parallel* to $NA$. If $P$ moves along $l$ in the opposite sense, $OP$ will tend to another limiting position, $OM$, and $OM \parallel NB$.

In EUCLIDEAN GEOMETRY, the two rays $OL$ and $OM$ form one line, and the angles $NOL$ and $NOM$ are right angles.

The hypothesis of HYPERBOLIC GEOMETRY is that the rays $OL$, $OM$ are distinct, so that Playfair's axiom is contradicted.

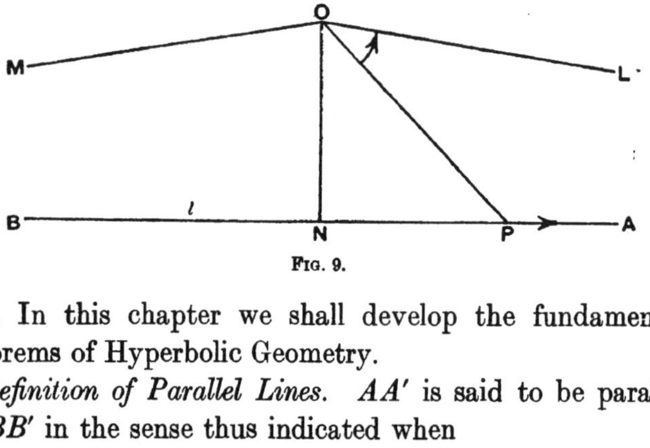


FIG. 9.

3. In this chapter we shall develop the fundamental theorems of Hyperbolic Geometry.

*Definition of Parallel Lines.* $AA'$ is said to be parallel to $BB'$ in the sense thus indicated when

(1) $AA'$ and $BB'$ lie in the same plane,

(2) $AA'$ does not meet $BB'$, both being produced indefinitely, and

(3) every ray drawn through $A$ within the angle $BAA'$ meets the ray $BB'$.

Through any point $O$ two parallels $OL$ and $OM$ can be drawn to a given line $AB$, so that $OL \parallel NA$ and $OM \parallel NB$. The angles $NOL$ and $NOM$ are, by symmetry, equal, and this angle depends only on the length of the perpendicular $ON = p$. It is called the *angle of parallelism* or the *parallel-angle*, and is denoted by $\Pi(p)$. There are two distinct senses of parallelism.

The two parallel lines separate all the lines through $O$ into two classes, those which intersect $AB$ and those which are non-intersectors of $AB$.

## Properties of Plane Figures, Parallelism, etc.

**4.** Parallel lines possess in common with euclidean parallels the following properties :

(1) *The property of parallelism is maintained, in the same sense, throughout the whole length of the line.* (Property of *transmissibility*.)

Let $AA' \parallel BB'$, and let $P$ be any point in $AA'$. We have

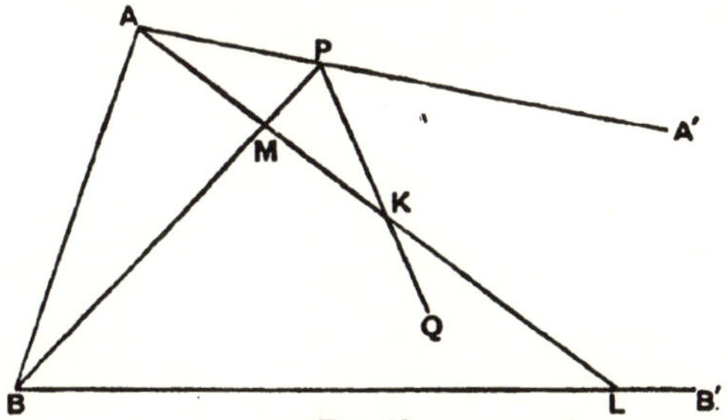


FIG. 10.

to prove that within the angle $BPA'$ every ray through $P$ cuts $BB'$, and no other ray through $P$ cuts $BB'$.

There are two cases to be considered, according as $P$ is on the side of $A$ in the direction of parallelism or not.

In the first case draw any line $PQ$ through $P$ within the angle $BPA'$, and on it take a point $K$. Then the line $AK$ must cut $BB'$ in some point $L$, and $BP$ in $M$. Hence $PQ$, which cannot cut again either $ML$ or $BM$, must cut the third side $BL$ of the triangle $BML$ (Pasch's Axiom). But $PA'$ does not cut $BB'$ ; therefore $PA' \parallel BB'$.

In the second case it is only necessary to take $K$ on $QP$ produced backwards.

(2) *Parallelism is reciprocal, i.e.* if $AA' \parallel BB'$, then $BB' \parallel AA'$ (Fig. 11).

The bisectors of the angles $BAA'$, $ABB'$ meet in a point $M$, since each meets the other parallel. Draw perpendiculars $MP$, $MQ$, $MR$ from $M$ on $AA'$, $BB'$, $AB$. By a comparison of the triangles these perpendiculars are equal. Draw $MM'$ bisecting the angle $PMQ$. Then, if $PQ$ is joined, $PQ \perp MM'$, and makes equal angles with $AA'$ and $BB'$. The lines $AA'$, $BB'$ are therefore symmetrical

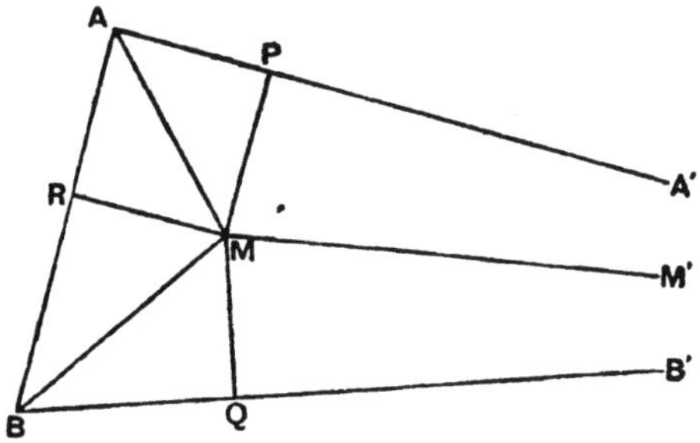


FIG. 11.

with respect to $MM'$, and the reciprocity is therefore established.

$P$, $Q$ are called *corresponding points* on the two parallels.

(3) *Parallelism is transitive, i.e.* if $AA' \parallel BB'$ and $BB' \parallel CC'$, then $AA' \parallel CC'$. There are two cases to be considered.

(*a*) Let $BB'$ lie between $AA'$ and $CC'$ (Fig. 12). We may suppose $ABC$ to be collinear. Within the angle $CAA'$ draw any line $AP$. Since $AA' \parallel BB'$, $AP$ cuts $BB'$ in a point $Q$. Then, since $QB' \parallel CC'$, $PQ$ produced within the angle $CQB'$ must cut $CC'$. Also $AA'$ itself does not cut $CC'$; therefore $AA' \parallel CC'$.

(*b*) In the same figure let $AA'$ and $BB'$ be $\parallel CC'$. Then any line within the angle $CAA'$ must cut $CC'$, and therefore $BB'$. Also $AA'$ itself cannot cut $BB'$, for then we would have two intersecting straight lines $AA'$ and $BB'$ both parallel to $CC'$ in the given sense. Therefore $AA' \parallel BB'$.

Parallels in hyperbolic geometry are, however, sharply distinguished from euclidean parallels by the following property :

The distance between two parallels diminishes in the direction of parallelism and tends to zero ; in the other direction the distance increases and tends to infinity.

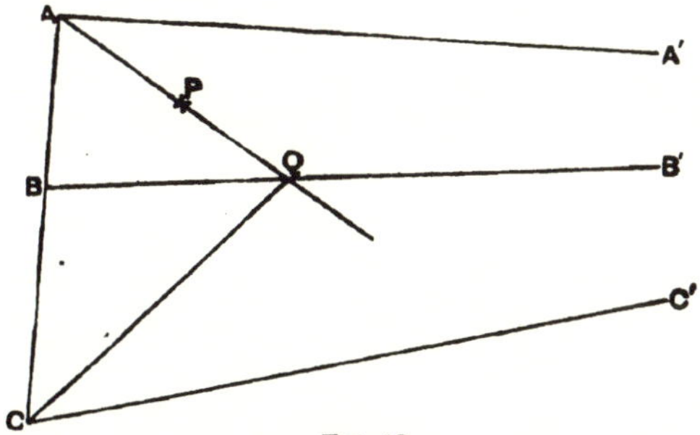


FIG. 12.

Before we can prove this we shall require several preliminary theorems.

5. *If a transversal meets two lines making the sum of the interior angles on the same side equal to two right angles, the two lines cannot meet and are not parallel.*

Let $PQ$ be a transversal cutting the two lines $AA'$ and $BB'$ in $P$ and $Q$ (Fig. 13), and making the sum of the angles $APQ + PQB$ equal to two right angles. Then, since the sum of the angles $PQB + B'QP = \pi$, therefore the alternate angles $APQ$ and $B'QP$ are equal.

Bisect $PQ$ at $M$ and draw $MK \perp AA'$ and $ML \perp BB'$.

Then the triangles $MKP$ and $MLQ$ are congruent and $\angle KMP = \angle LMQ$. Therefore $KML$ is a straight line, perpendicular to both $AA'$ and $BB'$.

By symmetry, if $AA'$ and $BB'$ meet on one side, they will also meet on the other. This is only possible in elliptic geometry. Also if $AA'$ and $BB'$ are parallel in one sense, they will be parallel also in the opposite sense, which is only true in euclidean geometry. Hence, in hyperbolic geometry they neither intersect nor are parallel.

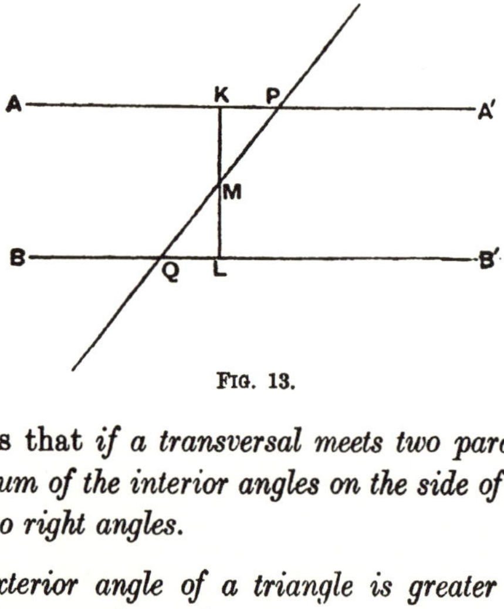


FIG. 13.

It follows that *if a transversal meets two parallel lines it makes the sum of the interior angles on the side of parallelism less than two right angles.*

**6.** *An exterior angle of a triangle is greater than either of the interior opposite angles.*

Let $ABC$ (Fig. 14) be a triangle with $BC$ produced to $D$. Then if the exterior angle $ACD$ is not greater than the interior angle $ABC$ it will be either equal to it or less.

Suppose first that

$$\angle ACD = ABC, \quad \text{then } \angle ACB + ABC = \pi$$

and $BA$, $CA$ cannot meet (except in elliptic geometry). Second. if $\angle ACD < ABC$, draw $BA'$ making $\angle A'BC = ACD$.

Then $BA'$ lies within the angle $ABC$ and must meet $AC$, while the sum of the angles $A'BC + A'CB = \pi$. But this is impossible (except in elliptic geometry).

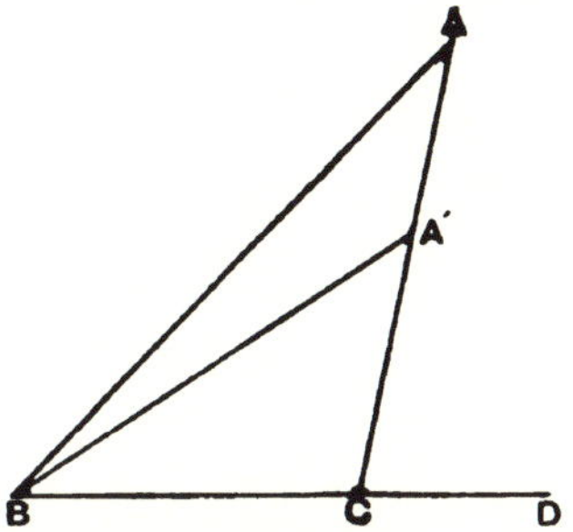


FIG. 14.

Hence the "theorem of the exterior angle" is true, except possibly in elliptic geometry.

7. *The parallel-angle* $\Pi(p)$ *diminishes as the distance* $p$ *increases.*

Let $AA'$ and $BB'$ be $\parallel MM'$ (Fig. 15), and $ABM \perp MM'$; and suppose $AM > BM$.

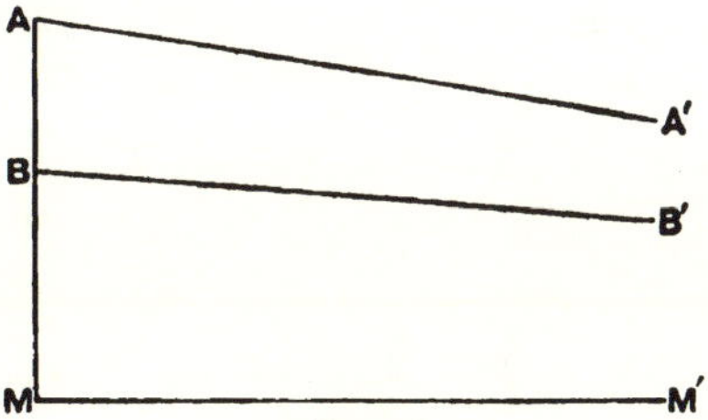


FIG. 15.

Then $\angle MAA' + ABB' < \pi$. (§ 5, *Cor.*)

But $\angle MBB' + ABB' = \pi$.

Therefore $\angle MAA' < MBB'$.

To avoid further prolixity we shall assume, or leave as exercises to the reader, the theorems that $\Pi(p)$ is uniquely defined for any value of $p$, and that there is a unique value of $p$ corresponding to any acute angle as parallel-angle. Further $\Pi(p)$ is a continuous function of $p$. As $p \to \infty$, $\Pi(p) \to 0$, and as $p \to 0, \Pi(p) \to \frac{\pi}{2}$. The analytical expression for $\Pi(p)$ will be found later in § 27. The range of $p$ may be extended into the negative region. If we suppose the point $A$ to move to the other side of the line $MM'$, the angle $MAA'$ will become obtuse, and we have, in fact,

$$\Pi(-p) + \Pi(p) = \pi.$$

**8.** (*a*) Let $ABNM$ be a quadrilateral with right angles at the adjacent vertices $M$, $N$, and let $MA = NB$. If we

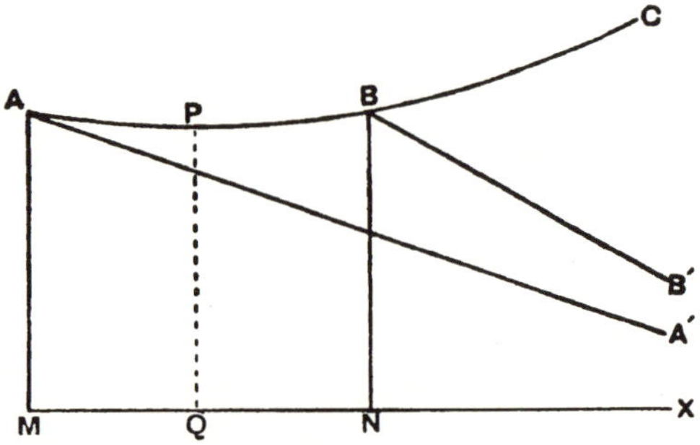


FIG. 16.

bisect $MN$ perpendicularly by $PQ$ we see from symmetry that the angles $MAB$ and $NBA$ are equal.

Draw $AA'$ and $BB' \parallel MN$. Then, since $MA = NB$, the angles $MAA'$ and $NBB'$ are equal.

But $\angle A'AB + B'BA < \pi$; therefore $\angle B'BC > A'AB$.

Hence $\angle MAB < NBC$, and the angles $MAB$ and $NBA$ are acute. Thus, hyperbolic geometry implies Saccheri's Hypothesis of the Acute Angle.

It follows, by considering the quadrilateral $AMQP$, that

*if a quadrilateral has three right angles the fourth angle must be acute.*

(*b*) *If $AM$, $BN$ are perpendiculars to $MN$ and $AM > BN$, then the angle $MAB < NBA$.*

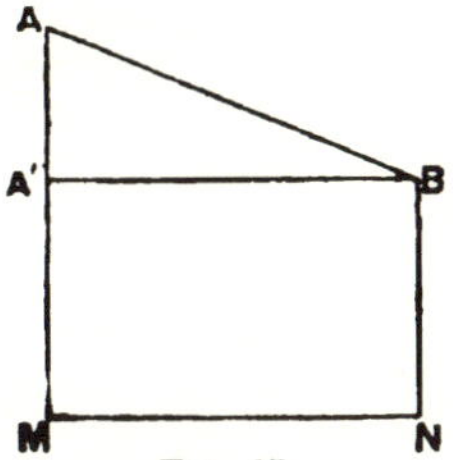


FIG. 17.

Cut off $MA' = NB$. Then

$$\angle NBA > NBA' = MA'B > MAB,$$

from the theorem of the exterior angle.

Conversely, if $\angle MAB < NBA$, then $MA > NB$. (Proof by *reductio ad absurdum*, using (*a*) and (*b*).)

9. *The distance between two intersecting lines increases without limit.*

Take two points $P$, $P'$ on $OA$ such that $OP' > OP$, and

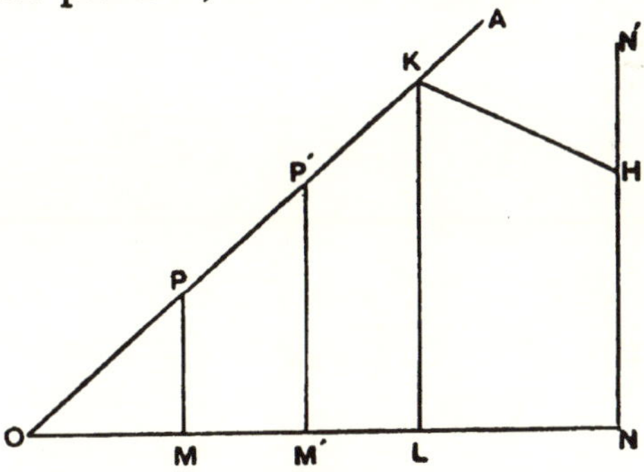


FIG. 18.

drop perpendiculars $PM$, $P'M'$ on $ON$. Then the angles $M'P'O$ and $MPO$ are both acute.

Therefore $\angle M'P'P < MPP'$, and hence $M'P' > MP$.

Take any length $G$. Let $ON$ be the distance corresponding to the parallel-angle $NOA$, and draw $NN' \perp ON$. Then $NN' \parallel OA$. Take $NH > G$, and draw a line $HK$ making the acute angle $N'HK$. Then $HK$, which lies within the angle $OHN'$, must meet $OA$ in some point $K$. Draw $KL \perp ON$. Then, since the angle $KHN$ is obtuse, $\angle LKH < NHK$; therefore $LK > NH > G$. Hence the perpendicular $PM$ can exceed any length.

**10.** (*a*) *The distance between two parallels diminishes in the direction of parallelism and tends to zero.*

Let $AA' \parallel MM'$, and let $AM$, $BN$ be two perpendiculars

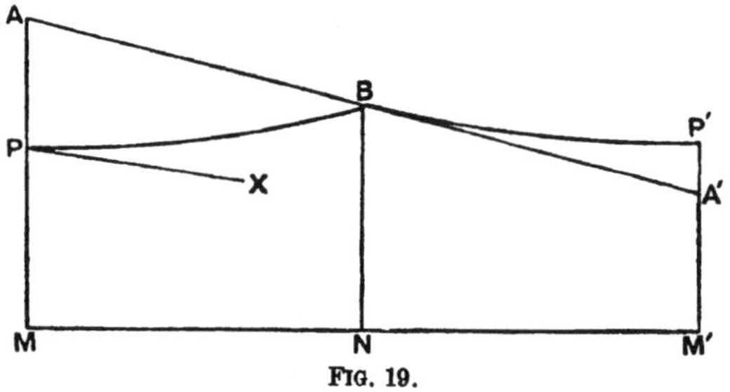


FIG. 19.

dropped on $MM'$ from points on $AA'$, such that $B$ lies on the side of $A$ in the direction of parallelism. Then the angles $MAA'$ and $NBA'$ are both acute; therefore $\angle MAB < NBA$, therefore $NB < MA$ (§ 8 (*b*), converse).

Choose any length $\epsilon$, however small, and make $MP < \epsilon$. Draw $PB \perp MA$. If $PX \parallel MM'$, $\angle MPX$ is acute; therefore $PB$ lies within the angle $APX$ and will meet $AA'$ in some point $B$, since $PX \parallel AA'$.

Make $\angle NBP' = NBP$, $BP' = BP$, and draw $P'M' \perp NM'$. Then $BP'$ neither meets nor is parallel to $NM'$, and $BA'$ must lie within the angle $M'BP'$, and therefore meets $M'P'$

in some point $A'$. Then $M'A' < M'P' < \epsilon$. Hence the distance between the parallels diminishes indefinitely.

Parallel lines are therefore *asymptotic*, and not equidistant as in euclidean geometry.

(*b*) *In the direction opposite to that of parallelism the distance between two parallels increases without limit.*

We have, in Fig. 19, $AM > BN$. Draw $AL \parallel M'M$ (Fig. 20). From $P$, any point on $A'A$, draw $PN \perp M'M$,

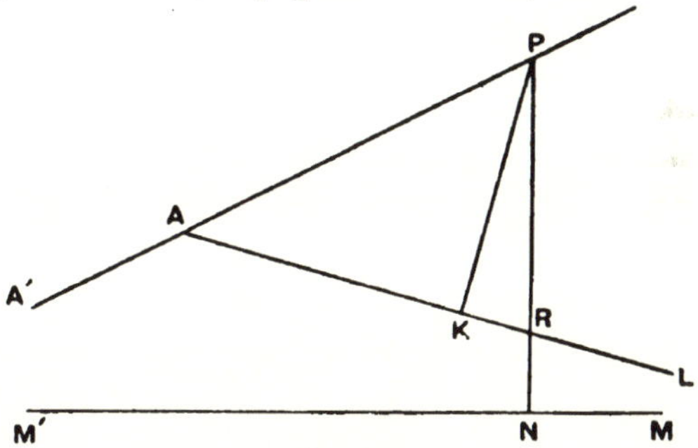


FIG. 20.

cutting $AL$ in $R$, and draw $PK \perp AL$. Then $PN > PR > PK$, and $PK$, the distance of $P$ from $AL$, can exceed any length. Hence $PN$ can exceed any length.

**11.** *Two parallel lines can therefore be regarded as meeting at infinity, and, further, the angle of intersection must be considered as being equal to zero.*

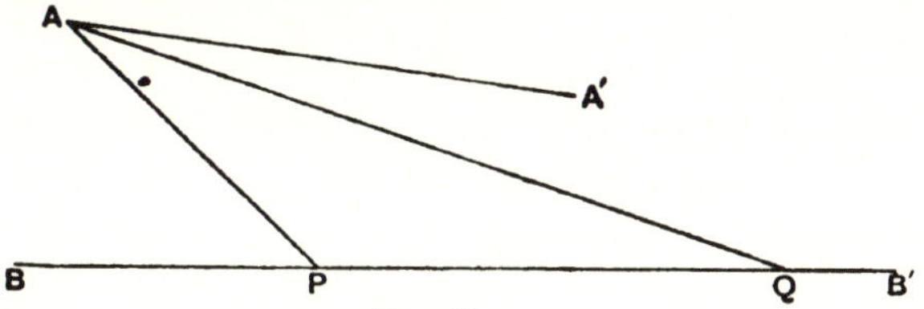


FIG. 21.

Let $AA' \parallel BB'$, and choose any small angle $\epsilon$. Draw

$AP$ making $\angle PAA' < \epsilon$. Then $AP$ cuts $BB'$ in some point $P$. Make $PQ = PA$, and join $AQ$. Then

$$\angle AQP = PAQ < PAA' < \epsilon.$$

Hence as $BQ \to \infty$, $AQ$ tends to the position $AA'$, and $\angle AQB \to 0$.

**12. Non-intersectors.**

If two lines are both perpendicular to a third, they cannot meet and are not parallel; and conversely, if two lines are non-intersecting and not parallel, they will have a common perpendicular.

Let $AB'$ and $LX$ be any two lines (Fig. 22). From any point $A$ on the one line draw a perpendicular $AL$ to the

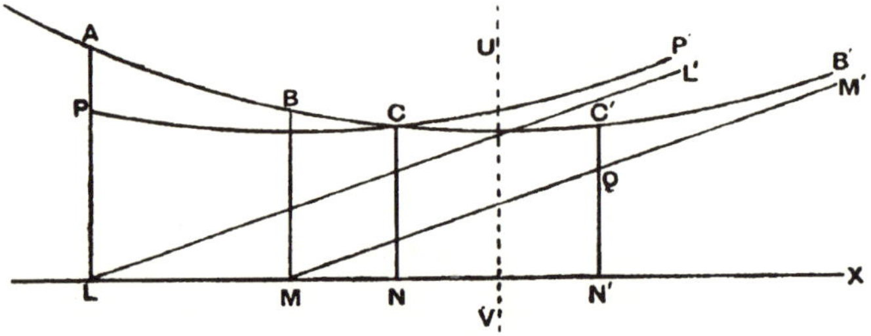


FIG. 22.

other. Then if $AL$ is not perpendicular to both lines it makes an acute angle with $AB'$ at one side, say the angle $LAB'$ is acute. If $BM$ is another perpendicular on the side of $AL$ next the acute angle, and such that the angle $MBB'$ is also acute, then $MB < LA$. The distance between the two lines thus diminishes in this direction, but unless the lines intersect or are parallel, it cannot diminish indefinitely.

Draw $MM' \parallel BB'$, and let the perpendicular $C'N'$ to $LX$ from any point $C'$ on $BB'$ meet $MM'$ in $Q$. Then

$C'N' > QN'$. But $QN' \to \infty$; therefore $C'N' \to \infty$. Thus the distance between the two given lines $AB'$ and $LX$ at first diminishes and finally tends to infinity. It must, therefore, have a minimum value, and, if $UV$ is that minimum distance, $UV$ must be perpendicular to both lines.

*Hilbert's*[1] *construction for the common perpendicular.*

Take $A$, $B$, any two points on the one line (Fig. 22), and draw perpendiculars $AL$, $BM$ on the other. If $AL = BM$, the common perpendicular is found by bisecting $LM$ perpendicularly.

Suppose $AL > BM$. Make $LP = MB$, and the angle $LPP' = MBB'$. Draw $LL' \parallel PP'$ and $MM' \parallel BB'$. Then, from the congruence of the figures $XLPP'$ and $XMBB'$, the angles $XLL'$ and $XMM'$ are equal. Hence $LL'$ is not parallel to $MM'$, and therefore is not parallel to $BB'$; nor does it meet $MM'$, therefore it must cut $BB'$. Therefore, since $PP' \parallel LL'$, it must meet $BB'$ in some point $C$. Make $BC' = PC$. Draw the perpendiculars $CN$, $C'N'$ to $LX$. Then, comparing $MBC'N'$ and $LPCN$, we find $CN = C'N'$, and the common perpendicular is found by bisecting $NN'$ perpendicularly by $UV$.

**13.** If we make the common perpendicular to two non-intersecting lines zero, the two lines will coincide, but if the common perpendicular at the same time goes off to infinity the two lines may become parallel.

Two straight lines may therefore be—

(1) *Intersecting*, and have a real angle of intersection, but no common perpendicular.

(2) *Non-intersecting*, and have a real shortest distance or common perpendicular, but no real angle.

[1] *Grundlagen der Geometrie*, 2nd ed. (1903), Appendix III. § 1.

(3) *Parallel*, with a zero angle and zero shortest distance or common perpendicular at infinity.

Before the principles of non-euclidean geometry became known, lines were sometimes classified as convergent, divergent and equidistant. In fact, from the assumption that two straight lines cannot first converge and then diverge without intersecting, Robert Simson (1756) was enabled to prove Euclid's postulate. In non-euclidean geometry equidistant straight lines cannot exist. Intersectors are convergent or divergent in the same sense as in euclidean geometry; parallels are convergent and asymptotic in one direction and divergent in the other; non-intersectors are ultimately divergent in both directions.

## Planes, Dihedral Angles, etc.

**14.** If two planes have a point in common they have a line in common.[1] The *dihedral angle* between two planes is measured in the usual way by the angle between two intersecting lines, one in each plane, perpendicular to the line of intersection. If this angle is a right angle the planes are perpendicular.

The usual proof in euclidean geometry that the dihedral angle measured in this way is independent of the point chosen on the line of intersection involves parallels, and another proof is required.

Take $P$, $P'$, any two points on the line of intersection of two planes $\alpha$, $\beta$ (Fig. 23). Draw $PA = P'A' \perp PP'$ in the plane $\alpha$, and $PB = P'B' \perp PP'$ in the plane $\beta$. Join $PA'$ and $P'A$ intersecting in $U$, and $PB'$ and $P'B$ intersecting in $V$. Then, by comparing the triangles $PAP'$ and $P'A'P$, we find $PA' = P'A$ and $\angle PAU = P'A'U$. Hence $PU = P'U$. Similarly $PB' = P'B$ and $PV = P'V$. Hence, by comparing triangles $PUV$, $P'UV$, we find $\angle UPV = UP'V$. Then, by comparing triangles $PA'B'$ and $P'AB$, we find $AB = A'B'$. Lastly, by comparing triangles $APB$ and $A'P'B'$, we obtain $\angle APB = A'P'B'$.

For the following theorems the usual proofs are valid.

[1] This is an *assumption*, explicitly excluding space of four or more dimensions.

If a straight line $p$ is perpendicular to each of two intersecting lines $a$, $b$ at their point of intersection $O$, it is perpendicular to every line through $O$ in the plane $ab$, and is said to be perpendicular to the plane $ab$. Every plane through $p$ is perpendicular to the plane $ab$. The line of

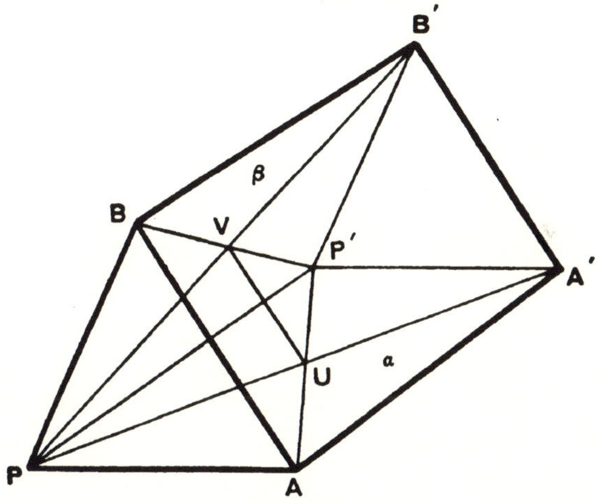


FIG. 23.

intersection of two planes which are both perpendicular to a plane $\alpha$ is perpendicular to $\alpha$. Two parallel lines lie in the same plane (by definition). Two lines $a$, $b$, which are both perpendicular to a plane $\gamma$, lie in a plane, for if $a$, $b$ cut $\gamma$ in $A$, $B$, then the planes $aB$ and $bA$ are both perpendicular to $\gamma$, and therefore coincide.

Three planes which have a point in common intersect in pairs in three concurrent straight lines. Three lines which intersect in pairs are either concurrent or coplanar.

**15.** (*a*) *If two lines $AA'$ and $CC'$ are both parallel to a third line $BB'$, then $AA' \parallel CC'$* (Fig. 24).

(The case in which all three lines lie in the same plane

has already been proved in § 4.) Take three fixed points *A*, *B*, *C* on the three lines, and any other point *P* on *BB'*.

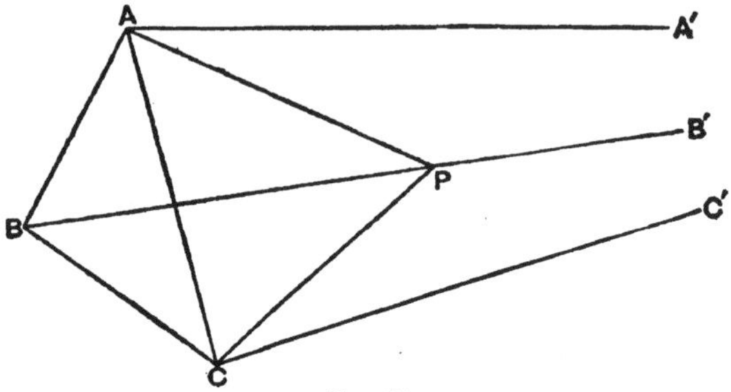


FIG. 24.

Join *PA*, *PC*. As *P* moves along *BB'*, the plane *PAC* rotates about *AC*. In the limit, *AP* and *CP* become parallel to *BB'*, and coincide respectively with *AA'* and *CC'*; therefore *AA'* and *CC'* lie in the same plane.

Again, if *CP* is fixed while the plane *PAC* revolves, *PA* tends to *PB'* and the plane *CPA* to *CPB'*. *CA*, the line

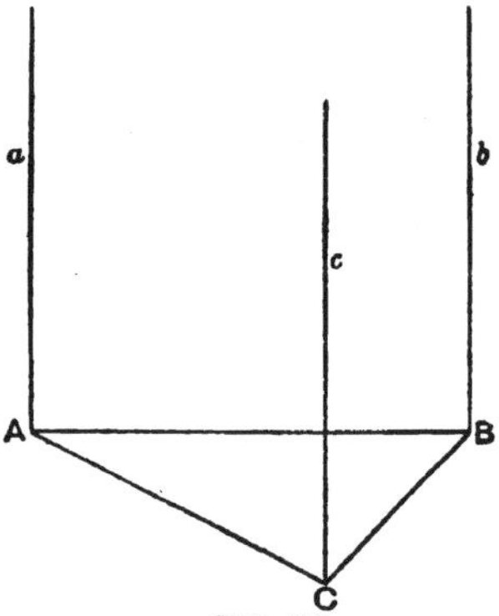


FIG. 25.

of intersection of the planes *CPA* and *C'CAA'*, therefore tends to *CC'*, and $CC' \parallel AA'$.

This result may be stated also in the form: *two planes*

*which pass respectively through two parallel lines intersect in a line parallel to the two given lines.*

(*b*) *If three planes* α, β, γ *intersect in lines* $a$, $b$, $c$, *such that* $a$ *and* $b$ *are neither parallel nor intersect, then* $a$, $b$, $c$ *are all perpendicular to the same plane* (Fig. 25).

Let $AB$ be the common perpendicular to $a$, $b$. Then the plane through $A \perp a$ passes through $B$ and is $\perp$ the plane $ab$, and therefore $\perp b$. Let this plane cut $c$ in $C$. Then the planes $ac$ and $bc$ are both perpendicular to the plane $ABC$, and therefore $a$, $b$, $c$ are all perpendicular to this plane.

**16. Pencils and bundles of lines.**

A system of coplanar lines through a point $O$ is called a *pencil* of lines with vertex $O$. The whole system of lines and planes through $O$ in space is called a *bundle* of lines and planes.

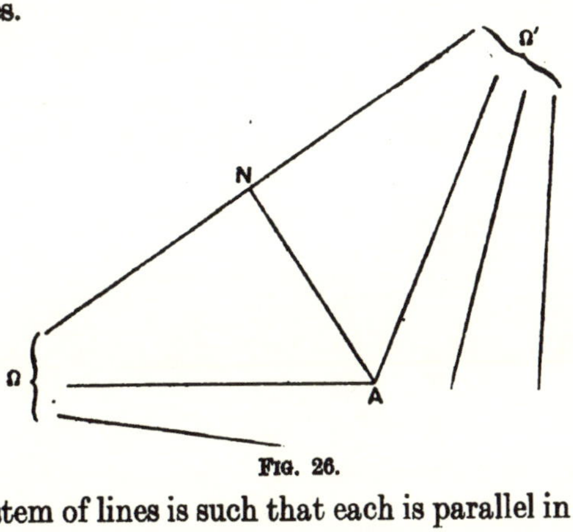


FIG. 26.

If a system of lines is such that each is parallel in the same sense to a given line, they are all parallel in pairs (§ 15 (*a*)), and form a pencil or bundle of parallel lines, or a *parallel bundle*. This is completely determined by one line with a given direction.

Denote a bundle of lines with vertex $O$ by $O$, and a bundle of lines parallel to $l$ in a given sense by $\Omega$. Then these two bundles uniquely determine a line $O\Omega$, which passes through $O$ and is parallel to $l$ in the given sense.

Two parallel bundles $\Omega$, $\Omega'$ uniquely determine a line $\Omega\Omega'$, which is parallel to both $l$ and $l'$. The line $\Omega\Omega'$ may be constructed thus: Take any point $A$ and determine $A\Omega$ and $A\Omega'$ (Fig. 26). Bisect the angle $\Omega A\Omega'$, and take the distance $AN$ corresponding to the parallel-angle $\frac{1}{2}\Omega A\Omega'$. The line through $N \perp AN$ in the plane $\Omega A\Omega'$ is $\parallel A\Omega$ and to $A\Omega'$.

So also we can prove that any three bundles, ordinary or parallel, uniquely determine a plane, for each pair determines a straight line, and the three straight lines thus determined are coplanar.

### 17. Points at infinity.

To an ordinary bundle corresponds a point $O$, but to a parallel bundle there is only a direction corresponding. We shall extend the class of points by introducing a class of fictitious points called *points at infinity*. These points function in exactly the same way as ordinary or, as we shall call them, *actual* points, and determine lines and planes with each other or with actual points.

On every line there are two points at infinity, and the assemblage of points at infinity in a plane is a curve of the second degree or conic,[1] since it is met by any line in two points. In three dimensions the assemblage is a surface of the second degree or quadric. This figure, the assemblage of all the points at infinity, is called the *Absolute.*

[1] The definition of a *conic* which we shall use is "a plane curve which is cut by any straight line in its plane in two points." For the explanation of the case of "imaginary" intersection see Chap. III. § 5.

When two points at infinity approach coincidence, the line determined by them becomes a tangent to the absolute. As $\Omega$, $\Omega'$ approach, the angle $\Omega A \Omega'$ in Fig. 26 tends to zero and $AN \to \infty$. The line $\Omega\Omega'$ therefore goes off to infinity. Such a line is called a *line at infinity*. Similarly we obtain *planes at infinity*, which are tangent planes to the Absolute.

In euclidean geometry there is just one parallel through a given point to a given line in a plane, and the two points at infinity upon a line coincide. The assemblage of points at infinity in a plane then reduces to a double line, the line at infinity, which is a degenerate case of a conic. There is in this case only one *real* line at infinity; but any line whose equation in rectangular coordinates is of the form $x \pm iy + c = 0$ is at an infinite distance from the origin, since $1 + i^2 = 0$, and the assemblage of these lines consists of two imaginary pencils. The equation of the line at infinity is $a \equiv 0x + 0y + 1 = 0$, and the equations of the two pencils are $\omega + \lambda a = 0$, $\omega' + \lambda a = 0$, where $\omega$, $\omega' = x \pm iy$.

The absolute in euclidean geometry thus consists, as a locus of points, of the line at infinity $a = 0$ taken twice, and, as an envelope of lines, of two imaginary pencils $\omega + \lambda a = 0$, $\omega' + \lambda a = 0$, with their vertices on the line at infinity. These two imaginary vertices are the points of intersection of the point-circle $\omega\omega' \equiv x^2 + y^2 = 0$ with the line at infinity. Since the equation of any circle can be written $\omega\omega' + ua = 0$, where $u = 0$ represents a straight line, we see that every circle passes through the two points ($\omega\omega' = 0$, $a = 0$), and for this reason these two imaginary points are called the *circular points*.

In euclidean geometry of three dimensions the absolute consists, as a locus of points, of the plane at infinity taken twice, and, as an envelope of planes, of all the planes through tangents to an imaginary circle, the intersection of the point-sphere $x^2 + y^2 + z^2 = 0$ with the plane at infinity.

The whole of metrical geometry is determined by the form of the Absolute; this will be more fully treated in Chap. V.

## 18. Ideal points.

*If a system of lines is such that any two are coplanar, while they do not all lie in the same plane and are neither parallel nor intersect, then they are all perpendicular to a fixed plane.*

If $a$, $b$ are any two of the lines they determine a plane $\pi$, which is perpendicular to both. If $c$ is any third line, which does not lie in the plane $ab$, it is the intersection of two planes $ac$ and $bc$, which are both $\perp \pi$, and therefore $c$ is $\perp \pi$ (§ 15 ($b$)).

We shall call this system, which is completely determined by two of the lines, or by a certain plane $\pi$, a bundle of lines with an *ideal* vertex **O**. The plane $\pi$ is called the axis of the bundle. All those lines of the system which lie in a plane are perpendicular to a straight line $l$, the intersection of their plane with the fixed plane, and form a pencil of lines with ideal vertex **O**. The line $l$ is called the axis of the pencil.

The *ideal points* thus introduced behave exactly like actual points. They can be regarded as lying outside the absolute, and are therefore ultra-spatial or ultra-infinite points.

Two ideal points may determine a real or actual line. Considering only points in a plane, the two ideal points are determined by two lines $a$, $a'$. If $a$, $a'$ are non-intersecting, the common perpendicular to these lines belongs to both pencils, and is therefore the line determined by the two ideal points. If $a \parallel a'$, the line **OO′** is a line at infinity; if $a$ cuts $a'$, **OO′** is an *ideal line*, which contains only ideal points. An ideal line lies entirely outside the absolute. Similarly, in three-dimensional hyperbolic geometry, we have *ideal planes*.

It is left to the reader to show now that any two points, actual, at infinity or ideal, always determine uniquely a line, actual, at infinity or ideal; and that any three points, actual, at infinity or ideal, always determine uniquely a plane, actual, at infinity or ideal.

These relations, in two dimensions, can be pictured more clearly if we draw a conic to represent the absolute, or assemblage of points at infinity (Fig. 26 *bis*). Actual points are then represented by points in the interior of the conic, ideal points by points outside the conic. Lines which intersect on the conic represent parallel lines, those

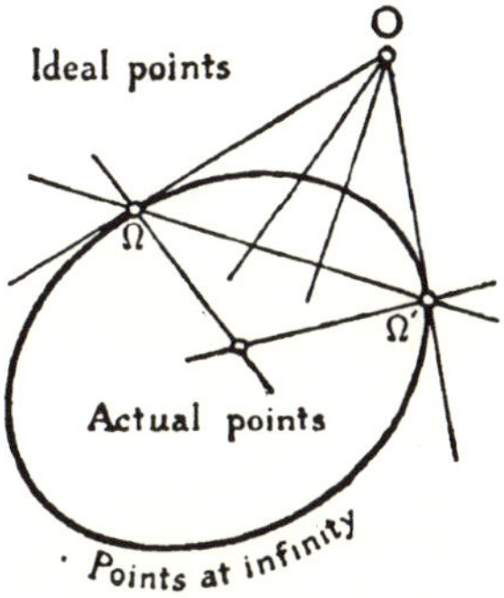


FIG. 26 *bis*.

which intersect outside the conic represent non-intersectors. If **O**Ω and **O**Ω′ are the tangents to the conic from an ideal point **O**, all the lines of the pencil with vertex **O** are perpendicular to ΩΩ′. For the present this may be used as a mere graphical representation. Its full meaning can only be understood in the light of projective geometry. (See Chap. III. §§ 5, 6, and Chap. V. §§ 1-14.)

### 19. Extension to three dimensions.

If the point of intersection of a line with a plane is at infinity, the line is said to be parallel to the plane. If the point of intersection is ideal there is a unique line and a unique plane perpendicular to both the given line and the given plane.

Two planes intersect in a line. If this line is at infinity the planes are said to be parallel; if it is ideal the two planes are non-intersecting and there is a unique line perpendicular to both.

All planes parallel to a given line in a given sense pass through the same point at infinity and intersect in pairs in a parallel bundle of lines.

All planes perpendicular to a given plane pass through the same ideal point and intersect in pairs in a bundle of lines with ideal vertex.

The following theorem is of great importance:

*Through a line which is parallel to a plane passes just one plane which is parallel to the given plane.*

Let the line $l$ cut the plane $\alpha$ in the point at infinity $\Omega$. Through $\Omega$ passes just one line at infinity $\omega$, and this line determines with $l$ a unique plane, which is parallel to $\alpha$. The actual construction may be obtained thus: Take any point $A$ on $l$ and draw $AN \perp \alpha$. Through $A$ draw $AB \perp$ the plane $lN$. Then $Bl$ is the plane required.

Through a line which meets a plane $\alpha$ in an ideal point O pass two planes parallel to the plane $\alpha$, for two tangents can be drawn from O to the section of the absolute made by the plane $\alpha$.

### 20. Principle of duality.

There is a correspondence between points and lines in a plane, and between points and planes in space, which gives rise to a sort of duality. To an actual plane $\alpha$ corresponds uniquely an ideal point $A$, all the lines and planes through which are perpendicular to the plane $\alpha$; and to an actual point $A$ corresponds an ideal plane $\alpha$, which is perpendicular to all the lines and planes through the point $A$. Let $B$ be any point on $\alpha$; then the plane $\beta$ which corresponds to $B$ must pass through $A$, since every plane perpendicular to $\alpha$ passes through $A$. If the plane $\alpha$ is at infinity the corresponding point $A$ is its point of contact with the absolute. The points and planes are therefore poles and polars with regard to the absolute. This reciprocity will appear again in elliptic geometry, where the elements are all real.

## The Circle and the Sphere.

### 21. The circle.

In a plane the locus of a point which is at a constant distance from a fixed point is a *circle*. The fixed point is the *centre*, and the constant distance the *radius*.

*A circle cuts all its radii at right angles.* This follows in the limit when we consider a chord $PQ$, which forms an isosceles triangle with the two radii $CP$, $CQ$. That is, a circle is the *orthogonal trajectory* of a pencil of lines with a real vertex.

Let the vertex go to infinity; then the lines of the pencil become parallel, and the circle takes a limiting form, which is not, as in ordinary geometry, a straight line, but is a uniform curve. This curve, a circle with infinite radius, is called a *horocycle*; it is the orthogonal trajectory of a pencil of parallel lines. The parallel lines, normal to the horocycle, are called its radii. All horocycles are superposable.

To obtain the orthogonal trajectory of a pencil of lines with ideal vertex we proceed thus:

Let $AA'$ be the axis of the pencil (Fig. 27), and draw perpendiculars to $AA'$. Cut off equal distances $MP$, $NQ$, ... along these perpendiculars. Then the locus of $P$ is again a uniform curve, which is not, as in ordinary geometry, a straight line; and the curve cuts all the perpendiculars to $AA'$ at right angles. It is therefore the orthogonal trajectory required. From the property that the curve is equidistant from the straight line $AA'$ it is called an *equi-*

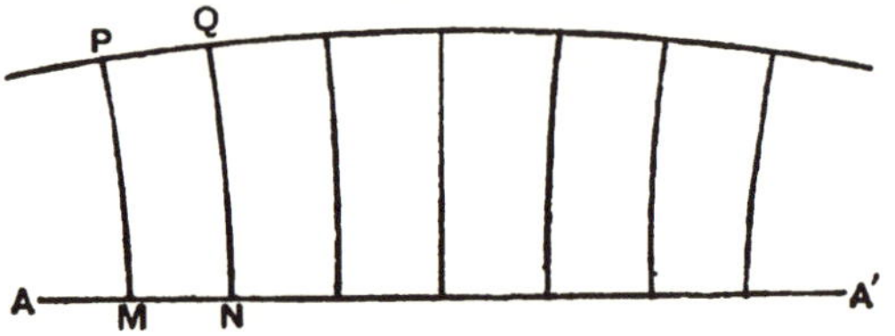


FIG. 27.

*distant-curve.* The complete curve consists of two branches, symmetrical about the axis, and also symmetrical about any line (radius) which is perpendicular to the axis.

As the axis tends to infinity, the perpendiculars tend to become parallel, and the equidistant-curve becomes a horocycle. We can thus pass continuously from an equidistant-curve to a circle. When the axis goes to infinity the centre also appears at infinity; then the axis becomes ideal and the centre becomes real.

There are therefore three sorts of circles:

(1) *Proper circles,* with real centre and ideal axis.

(2) *Horocycles,* with centre and axis at infinity.

(3) *Equidistant-curves,* with ideal centre and real axis.

A straight line, or rather two coincident lines, is the limiting case of an equidistant-curve when the distance vanishes.

## 22. The sphere.

In space of three dimensions the locus of a point which is equidistant from a fixed point is a sphere. It is the orthogonal trajectory of a bundle of lines with a real vertex. When the centre is at infinity the surface is called a *horo-*

*sphere*, and when the centre is ideal the surface is an *equidistant-surface* to a plane as axis.

A plane section of a sphere is always a circle; the greatest section, or the section of least curvature, is a diametral section passing through the centre, that is, a great circle on the sphere.

A plane section of a horosphere is a circle, except when the section is normal to the surface, *i.e.* passes through a normal, in which case the section is a horocycle.

A section of an equidistant surface by a plane which does not cut the axial plane is a circle; if the plane cuts the axial plane the section is an equidistant-curve with the line of intersection as axis; if the plane is parallel to the axial plane the section is a horocycle.

**23. Circles determined by three points or three tangents.**

Let $A$, $B$, $C$ be three given points: to find the centre of a circle passing through $A$, $B$, $C$. Bisect the joins of the three points

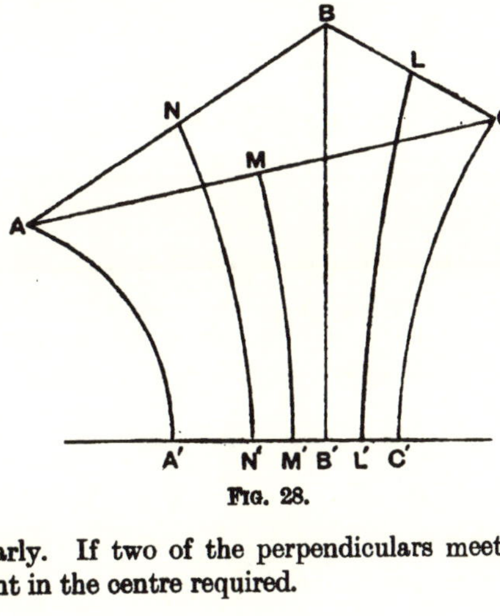


FIG. 28.

perpendicularly. If two of the perpendiculars meet, all three will be concurrent in the centre required.

Suppose the perpendicular bisectors $NN'$ and $LL'$ to $AB$ and $BC$ are non-intersecting. Let $N'L'$ be their common perpendicular. Let this line cut the perpendicular at $M$ in $M'$, and draw the perpendiculars $AA'$, $BB'$, $CC'$. Then, since $NN'$ bisects $AB$ perpendicularly and is $\perp A'B'$, $AA' = BB'$. Similarly $BB' = CC'$.

In the quadrilaterals $AA'M'M$, $CC'M'M$, $AA' = CC'$, $AM = MC$ and the angles at $M$, $A'$ and $C'$ are right; therefore if the quadrilateral $CC'M'M$ be folded over $MM'$, $C$ will coincide with $A$, and, since only one perpendicular can be drawn from $A$ upon $A'M'$, $CC'$ will coincide with $AA'$, and the angles at $M'$ are right. Hence $LL'$, $MM'$, $NN'$ are all perpendicular to $A'C'$, and $ABC$ lie on an equidistant-curve with axis $A'B'C'$.

Suppose $NN' \parallel LL'$; then $MM'$ must be parallel to both. For, if $MM'$ cuts $LL'$, then by the first case the three lines are concurrent; and if $MM'$ is a non-intersector to $LL'$, then by the second case $LL'$ and $NN'$ are non-intersecting. Therefore $MM' \parallel LL'$. $A$, $B$, $C$ then lie on a horocycle.

In addition to the circle, equidistant-curve or horocycle, which can be drawn through $ABC$ in this way, there exist three equidistant-curves such that one of the points lies on one branch while the other two lie on the other branch. Bisect $AB$, $AC$ at $M$ and $N$. Join $MN$ and draw the perpendiculars $AA'$, $BB'$, $CC'$ to $MN$. (See Fig. 50, p. 77.) Then $AA' = BB' = CC'$, and an equidistant-curve with axis $MN$ passes through $B$, $C$ and $A$. A triangle has therefore four circumcircles, at least three of which are equidistant-curves. There cannot be more than one real circumcentre. This point, which we may call *the* circumcentre, is the point of concurrence of the perpendicular bisectors of the sides, and may be real, at infinity, or ideal.

If $L$ is the middle point of $BC$, the perpendicular from $L$ on $MN$ is also $\perp BC$, since it bisects the quadrilateral $BB'C'C$. Hence the altitudes of the triangle $LMN$ are concurrent in the circumcentre of the triangle $ABC$. A triangle therefore possesses a unique *orthocentre*, real, at infinity, or ideal. If the orthocentre is ideal there is a real *orthaxis*, which is perpendicular to the three altitudes.

The construction for the circles touching the sides of a triangle is, as usual, obtained by bisecting the angles. Three of the circles may be equidistant-curves or horocycles.

**24. Geometry of a bundle of lines and planes.**

In plane geometry we have points, lines, distances and angles; in a bundle of lines and planes through a point $O$ we have lines, planes, plane angles and dihedral angles Let us change the language to make it resemble the language of plane geometry. In translating from one language to another we require a dictionary. The following will suffice:

| | |
|---|---|
| "Point" - - | Line through $O$. |
| "Line" - - - | Plane through $O$. |
| "Distance" between two "points" - | Angle between two lines through $O$. |
| "Angle" between two "lines" - | Dihedral angle between two planes through $O$. |
| "Parallel lines" - | Parallel planes. |

Then two "points" determine a "line" and two "lines" determine a "point." "Parallel lines" only exist when $O$ is at infinity or ideal.

When $O$ is at infinity, through a given "point" there passes just one "line" "parallel" to a given "line" (§ 19), and when $O$ is ideal, two "parallels" can be drawn through a given "point" to a given "line."

There are therefore three kinds of geometry of a bundle according as the vertex $O$ is actual, at infinity or ideal, and these are exactly of the same form as elliptic, parabolic (*i.e.* euclidean) and hyperbolic plane geometry.

If a sphere be drawn with centre $O$ cutting the lines and planes of the bundle, we can get a further correspondence. When $O$ is an actual point we have a proper sphere. We have then the following dictionary:

| | |
|---|---|
| "Point" - | Pair of antipodal points on sphere. |
| "Line" - - | Great circle. |

| | |
|---|---|
| "Distance" - - | Length of arc. |
| "Angle" between "lines" | Angle between great circles. |

Hence the geometry on a proper sphere, where great circles represent lines, and pairs of antipodal points represent points, is the same as elliptic geometry. (See further Chapter III.)

When $O$ is ideal, the sphere becomes an equidistant-surface, and its geometry is hyperbolic; when $O$ is at infinity it becomes a horosphere, and its geometry is euclidean: "point" in each case being represented by a point, and "lines" by normal sections, which are also shortest lines or geodesics on the surface.

We have here the important and remarkable theorem that *the geometry on the horosphere is euclidean.*

## Trigonometrical Formulae.

**25**. We shall now proceed to investigate the metrical relations of figures, leading up to the trigonometrical formulae for a triangle. The starting point is found in a relation connecting the arcs of concentric horocycles, and this leads to the expression for the angle of parallelism. The great theorem which enables us to introduce the circular functions, sines and cosines, etc., of an angle is that the geometry of shortest lines (horocycles) traced on a horosphere is the same as plane euclidean geometry. Let $A$, $B$, $C$ be three points on a horosphere with centre $\Omega$. The planes $AB\Omega$, etc., cut the surface in horocycles, and we have a triangle $ABC$ formed of shortest lines or geodesics, which are arcs of horocycles. The angles of this triangle are the angles between the tangents to the arcs or the dihedral angles between the planes $\Omega AB$, $\Omega BC$, $\Omega CA$. If the angle at $C$ is a right angle, then the ratios of the arcs are

$$\frac{BC}{AB}=\sin A, \quad \frac{AC}{AB}=\cos A, \text{ etc.}$$

The circular functions could be introduced independently of the horosphere by defining them as analytical functions of the angle $\theta$, viz.:

$$\sin\theta=\theta-\frac{\theta^3}{3!}+\frac{\theta^5}{5!}-\dots,$$

$$\cos\theta=1-\frac{\theta^2}{2!}+\frac{\theta^4}{4!}-\dots,$$

the unit of angle being such that the measure of a flat-angle is $\pi=3{\cdot}14159\dots$. We may call this "circular measure." Then it could be shown that if $ABC$ is a rectilinear triangle, right-angled at $C$, the ratios $BC/AB$ and $AC/AB$ tend to the limits $\sin A$ and $\cos A$ as $BC$, $AC$ and $AB$ all tend to zero, while the angle $A$ is fixed. (Cf. Chap. III. § 18, footnote).

### 26. Ratio of arcs of concentric horocycles.

Let $A_1\Omega$, $B_1\Omega$ be two parallel lines, and let them be cut by horocycles $A_1B_1$, $A_2B_2$, $A_3B_3$ with centre at infinity $\Omega$.

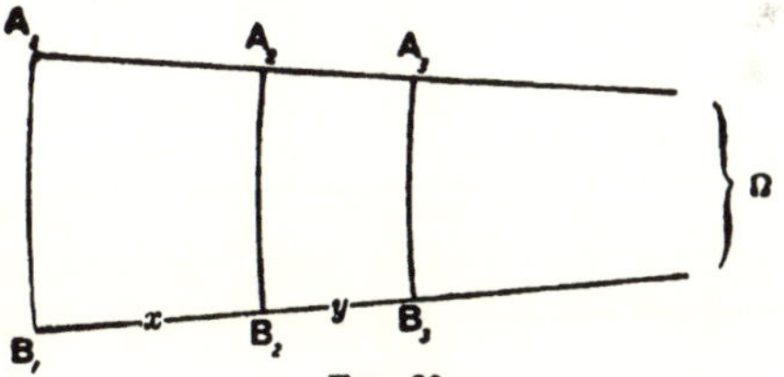


FIG. 29.

Then the ratio of the arcs $A_1B_1 : A_2B_2$ depends only on the distance $A_1A_2=x$. (See Ex. 25 and 26.)

Let $\frac{A_1B_1}{A_2B_2}=f(x)$; then $\frac{A_2B_2}{A_3B_3}=f(y)$ and $\frac{A_1B_1}{A_3B_3}=f(x+y)$.

Therefore $$f(x+y)=f(x).f(y).$$

This is the fundamental law of indices, and the function is therefore the exponential function :

$$f(x) = c^x,$$

$c$ being an absolute constant greater than unity. $f(x)$ is a pure ratio, and must be independent of the arbitrary unit of length which is selected ; therefore $\log f(x)$ or $x \log c$ must be a pure ratio. Hence $\log c$ must be the reciprocal of a length. We shall put $\log_e c = 1/k$ ; then

$$f(x) = e^{\frac{x}{k}},$$

where $k$ is an absolute linear constant and $e$ is the base of the natural logarithms. $k$ is called the space-constant ; its actual value in numbers of course depends upon the arbitrary unit of length which is selected, but it forms itself a natural unit of length, and it is often convenient to make its value unity. This is one of the most remarkable facts in non-euclidean geometry, that there is an absolute unit for length as well as for angle. It can be proved (see § 39) that *$k$ is the length of the arc of a horocycle which is such that the tangent at one extremity is parallel to the radius through the other extremity.*

**27. The parallel-angle.**

We can apply this now to find the value of the parallel-angle $\Pi(p)$ in terms of $p$. This is the simplest case of the determination of the relations between the sides and angles of a triangle. The triangle in this case has two sides infinite, one angle right and another angle zero.

Let $A\Gamma \parallel B\Gamma$, and $AB \perp B\Gamma$ (Fig. 30). Erect a perpendicular at $A$ to the plane of $AB\Gamma$. Draw the parallels $B\Omega$ and $\Gamma\Omega$. Draw the horosphere with centre at infinity $\Omega$ and passing through $A$, and let it cut $B\Omega$ in $B'$ and $\Gamma\Omega$ in $C'$. Let $BB' = y$, and the arcs $B'C'$, $C'A$, $AB'$ be $a$, $b$, $c$.

Since $B\Gamma \perp AB$ the plane $\Omega B\Gamma \perp$ plane $\Omega AB$, and since the angles which the arcs $AB'$, $B'C'$, $C'A$ make with the lines $\Omega A$, $\Omega B$, $\Omega C'$ are all right,

$$\angle C'AB' = \Gamma AB = \Pi(p), \quad \angle AB'C' = \frac{\pi}{2},$$

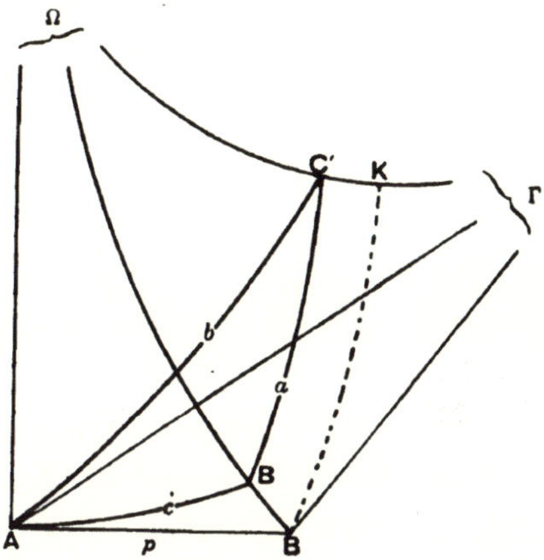


FIG. 30.

and since geometry on the horosphere is euclidean,

$$\angle AC'B' = \frac{\pi}{2} - \Pi(p).$$

Hence $\sin \Pi(p) = \frac{a}{b}$ and $\cos \Pi(p) = \frac{c}{b}$;

therefore $$\tan \tfrac{1}{2}\Pi(p) = \frac{a}{b+c}. \quad \text{.........................(1)}$$

The arc of the horocycle $b$ is a standard length, viz. the length of the arc which is such that the tangent at one extremity is parallel to the radius at the other extremity. In Fig. 30 $BK$ is such an arc, and $=b$. Hence

$$\frac{b}{a} = e^{\frac{y}{k}}. \quad \text{.................................(2)}$$

Now fold the plane $\Omega A\Gamma$ about $\Omega A$ until it lies in the plane $\Omega AB$ (Fig. 31). Draw $\Gamma B'' \perp B\Omega$. Then, if we draw

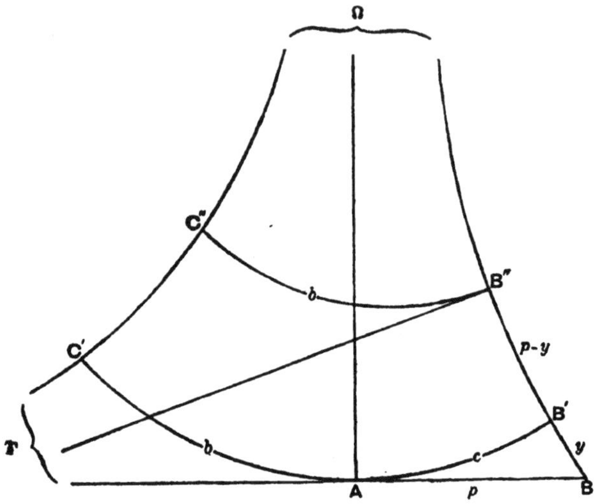


FIG. 31.

the horocyclic arc $B''C''$ with centre $\Omega$, this arc $= b$. Also $BB'' = p$; therefore $B'B'' = p - y$.

Therefore $$\frac{b+c}{b} = e^{\frac{p-y}{k}}. \qquad \text{(3)}$$

Hence, multiplying (2) and (3) and using (1), we have

$$\tan \tfrac{1}{2}\Pi(p) = e^{-\frac{p}{k}}.$$

This relation may be put into other forms, *e.g.*

$$\cot \Pi(p) = \sinh \frac{p}{k};$$

other equivalent forms can be read off from the accompanying figure (Fig. 32), treating the figure as a euclidean triangle.

We shall effect a simplification by taking in the following paragraphs (§§ 28-37) the constant $k$ as the unit of length.

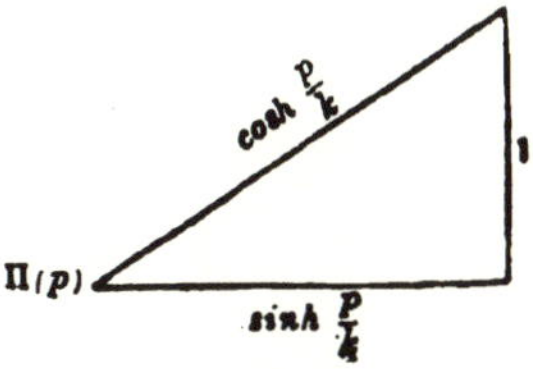


FIG. 32.

The formulae may be restored in their general form by dividing by $k$ every letter which represents a length.

## 28. Two formulae for the horocycle.

Let $AB = s$ be an arc of a horocycle with centre $\Omega$, and let $S$ be the length of an arc of a horocycle such that the tangent at one end is perpendicular to the radius at the other end (Fig. 33). Let $s < S$.

Extend $BA$ to $M$ so that $BM = S$; then $AM = S - s$. Take $A_1$ on the radius through $A$ so that the perpendicular at $A_1$ to $A_1A$ is parallel to $MM_1$. Then the arc $A_1M_1 = S$. Let $B\Omega_1$ cut $A_1\Omega$ in $D$. Then $DA_1 = DB = t$, say. Let $DA = u$. Then we have, comparing the arcs $A_1M_1$ and $AM$,

$$S - s = Se^{-t-u}. \qquad (1)$$

Extend $AB$ to $N$ so that $BN = S$. Take $A_2$ on the radius through $A$ so that the perpendicular at $A_2$ to $AA_2$ is parallel to $M_2N$. Then the arc $A_2M_2 = S$. And since $B\Omega \parallel DA_2$, and $A_2\Omega_2 \parallel DB$, $DA_2 = DB = t$.

Then

$$S + s = Se^{t-u}. \qquad (2)$$

Adding these two equations, we get

$$2S = Se^{-u}(e^t + e^{-t})$$

or

$$e^u = \tfrac{1}{2}(e^t + e^{-t}) = \cosh t. \qquad (3)$$

Substituting in (1), we get

$$s = S\left(1 - \frac{2e^{-t}}{e^{t}+e^{-t}}\right) = S\,\frac{e^{t}-e^{-t}}{e^{t}+e^{-t}} = S\tanh t. \quad \text{............(A)}$$

Draw $s'$ the arc of the horocycle with centre $\Omega$ passing through $D$.

Then $\qquad s' = se^{u} = S\tanh t \,.\, \cosh t = S\sinh t. \quad \text{..............(B)}$

These two formulae (A) and (B) give the tangent and ordinate at the extremity of an arc of a horocycle, viz. if

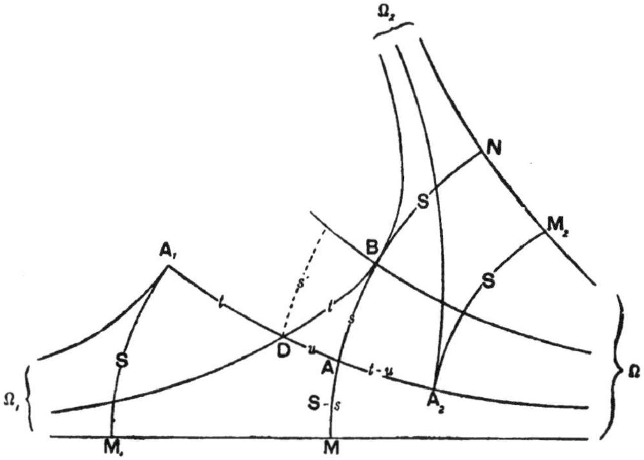


FIG. 33.

$s$ is the arc $AB$ of a horocycle, $t$ the length of the tangent $AT$ intercepted between the point of contact $A$ and the radius through $B$, and $y$ the ordinate $AN$ from $A$ on the radius $BT$, then

$$s = S\tanh t = S\sinh y.$$

## 29. The right-angled triangle; complementary angles and segments.

Let $ABC$ be a right-angled triangle with right angle at $C$. Denote the sides by $a$, $b$, the hypotenuse by $c$, the angles opposite the sides by $\lambda$, $\mu$.

Let $\alpha = \Pi(a)$, etc. Then we have five segments and five angles connected by the relations

$$\alpha = \Pi(a), \quad \beta = \Pi(b), \quad \gamma = \Pi(c), \quad \lambda = \Pi(l), \quad \mu = \Pi(m).$$

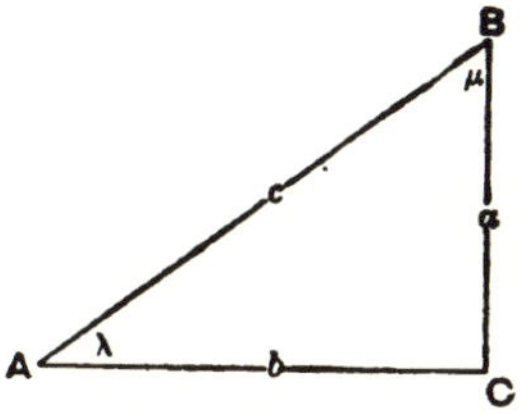


FIG. 34.

Let $a'$ denote $\frac{1}{2}\pi - a$; then we have the complementary segments and angles

$$\alpha' = \Pi(a'), \quad \beta' = \Pi(b'), \text{ etc.}$$

We have to deal with the circular functions of the angles, and the complementary angles are of course connected by the relations

$$\sin \alpha' = \cos \alpha, \quad \tan \alpha' = \cot \alpha, \text{ etc.}$$

We have also to deal with the hyperbolic functions of the segments, and we have the relations

$$\sinh a = \cot \Pi(a) = \cot \alpha = \tan \alpha' = \tan \Pi(a') = \operatorname{cosech} a',$$
$$\cosh a = \operatorname{cosec} \Pi(a) = \operatorname{cosec} \alpha = \sec \alpha' = \sec \Pi(a') = \coth a'.$$

## 30. Correspondence between rectilinear and spherical triangles.

Draw $A\Omega \perp$ the plane of the triangle (Fig. 35), and draw $B\Omega$ and $C\Omega \parallel A\Omega$.

Then $B\Omega \parallel C\Omega$, and $BC \perp$ plane $AC\Omega$; therefore $BC \perp C\Omega$. The plane $\Omega BA \perp$ plane $ABC$, and the angle between the planes $\Omega BC$ and $ABC = \Pi(b)$. Also, since the planes $\Omega AB$,

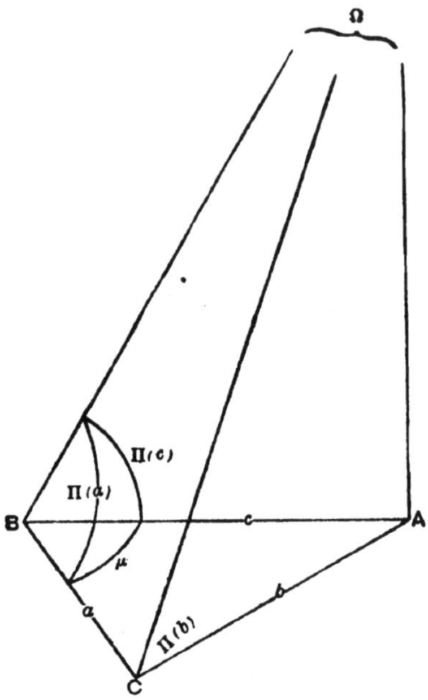


FIG. 35.

$\Omega BC$, $\Omega CA$ intersect in parallel lines, the sum of the angles of intersection $= \pi$; therefore the angle between the planes $\Omega AB$ and $\Omega BC = \frac{\pi}{2} - \lambda$.

Now draw a sphere with centre $B$, and we get a right-angled spherical triangle with hypotenuse $\alpha = \Pi(a)$, sides $\mu$ and $\gamma = \Pi(c)$, measured by the angles which they subtend at the centre, and opposite angles $\lambda' = \frac{\pi}{2} - \lambda$ and

$\beta = \Pi(b)$, *i.e.* to the rectilinear triangle $(c, a\lambda, b\mu)$ corresponds the spherical triangle $(a, \mu\lambda', \gamma\beta)$.

## 31. Associated triangles.

To the spherical triangle $(a, \mu\lambda', \gamma\beta)$ we get four other associated triangles by drawing the polars of the two vertices (cf. Chapter III. § 20). This gives a star pentagon (Fig. 36) whose outer angles are all right angles. The five

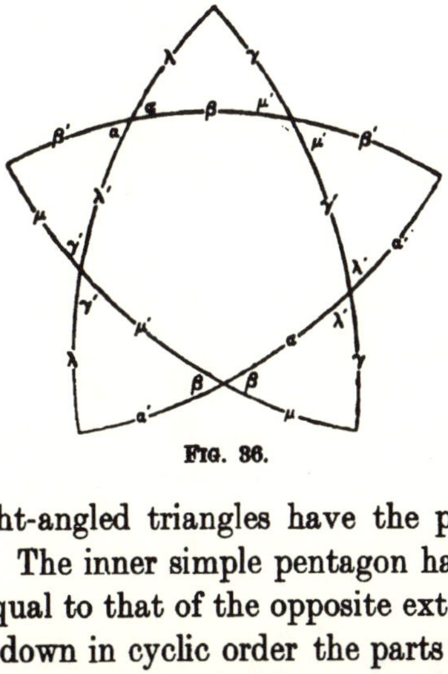

FIG. 36.

associated right-angled triangles have the parts indicated in the figure. The inner simple pentagon has the measure of each side equal to that of the opposite exterior angle.

If we write down in cyclic order the parts $\lambda', \mu', a, \gamma', \beta$ as they occur on the sides of the simple pentagon, the parts of the five associated spherical triangles can be written down by cyclic permutation of these letters, thus:

1. $a, \quad \mu\ \lambda', \quad \gamma\ \beta.$
2. $\gamma', \quad a'\ \mu', \quad \beta'\ \lambda'.$
3. $\beta, \quad \gamma\ a, \quad \lambda\ \mu'.$
4. $\lambda', \quad \beta'\ \gamma', \quad \mu\ a.$
5. $\mu', \quad \lambda\ \beta, \quad a'\ \gamma'.$

Corresponding to these we get five associated rectilinear triangles :

| | | | |
|---|---|---|---|
| 1. | $c$ , | $a$ $\lambda$ , | $b$ $\mu$ . |
| 2. | $b'$ , | $c'$ $\mu$ , | $l'$ $a'$ . |
| 3. | $l$ , | $b$ $a'$ , | $m'$ $\gamma$ . |
| 4. | $m$ , | $l'$ $\gamma$ , | $a$ $\beta'$ . |
| 5. | $a'$ , | $m'$ $\beta'$ , | $c'$ $\lambda$ . |

Hence, if we establish a relationship between the sides and angles of one triangle, we can obtain four other relationships by applying the same result to the associated triangles, or by a cyclic permutation of the letters $(l'm'ac'b)$ $(lma'cb')$ $(\lambda'\mu'a\gamma'\beta)$ $(\lambda\mu a'\gamma\beta')$.

## 32. Trigonometrical formulae for a right-angled triangle.

Produce the hypotenuse $AB$ to $D$ so that the perpendicular at $D$ to $AD$ is parallel to $AC$. Then $AD=l$,

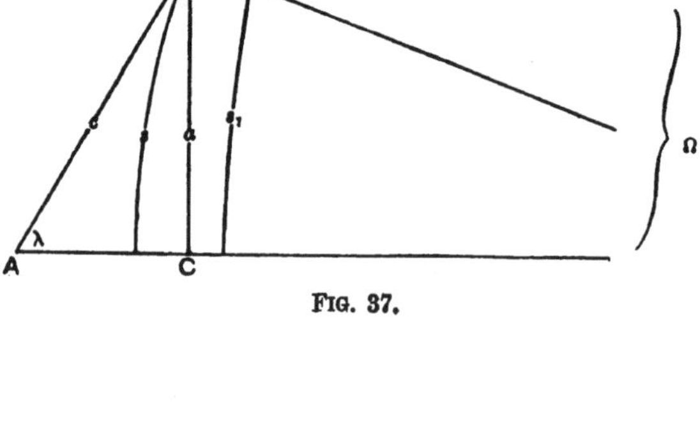


FIG. 37.

$BD = l - c$. Draw the horocyclic arcs with centre $\Omega$ passing through $D$ and $B$. Then (by § 28 (A), (B) and (3).)

$$s_1 + s_2 = S \tanh l, \qquad s_2 = S \tanh (l - c),$$
$$s_1 = se^{-u} = S \sinh a / \cosh (l - c).$$

Therefore

$$\tanh l = \tanh (l - c) + \sinh a / \cosh (l - c),$$
$$\sinh a \cosh l = \sinh l \cosh (l - c) - \cosh l \sinh (l - c) = \sinh c,$$

or $$\sinh a = \sinh c \sin \lambda. \qquad \text{(1)}$$

From the associated triangles we get

| | | |
|---|---|---|
| $\sinh c' = \sinh b' \sin \mu$; therefore | $\sinh b = \sinh c \sin \mu.$ | (2) |
| $\sinh b = \sinh l \sin \alpha'$ | $\sinh b = \tanh a \cot \lambda.$ | (3) |
| $\sinh l' = \sinh m \sin \gamma$ | $\cosh c = \cot \lambda \cot \mu.$ | (4) |
| $\sinh m' = \sinh a' \sin \beta'$ | $\sinh a = \tanh b \cot \mu.$ | (5) |
| From (3), (4) and (5) we get | $\cosh c = \cosh a \cosh b.$ | (6) |
| (1), (4) and (5) | $\cos \lambda = \tanh b \coth c.$ | (7) |
| (1), (2) and (3) | $\cos \lambda = \cosh a \sin \mu.$ | (8) |
| (1), (2) and (5) | $\cos \mu = \cosh b \sin \lambda.$ | (9) |
| (2), (3) and (4) | $\cos \mu = \tanh a \coth c.$ | (10) |

### 33. Engel-Napier rules.

These ten formulae, which connect all the five parts of the triangle in sets of three, are of exactly the same form as the formulae of spherical trigonometry, with hyperbolic

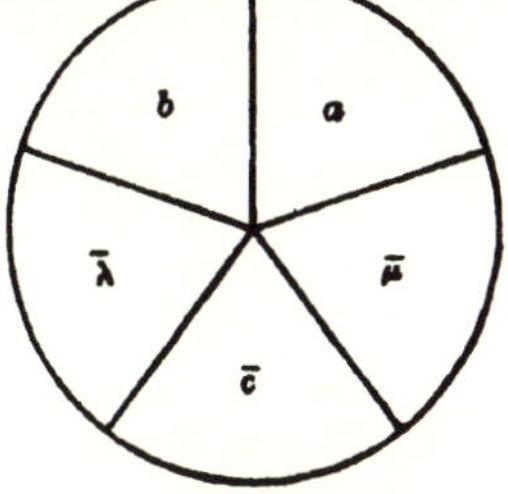


FIG. 38.

functions of the sides instead of circular functions, and can be written down by Napier's rules.[1] If we write the five parts $b$, $\bar{\lambda}$, $\bar{c}$, $\bar{\mu}$, $a$ in cyclic order as they occur in the triangle (Fig. 38), then

sine (middle part) = product of cosines of opposite parts

(A) $\qquad\qquad\qquad$ = product of tangents of adjacent parts,

it being understood that the circular functions of the angles, and the hyperbolic functions of the sides, are taken, and each function of $\bar{x}$ is the " complement " of the corresponding function of $x$, *i.e.* $\cosh \bar{c} = \sinh c$, $\tan \bar{\lambda} = \cot \lambda$, etc.

[NOTE.—$\bar{\lambda}$ has the same meaning as $\lambda'$, but $\bar{c}$ is not the same as $c'$.]

This rule may be put in another form, which is more homogeneous,

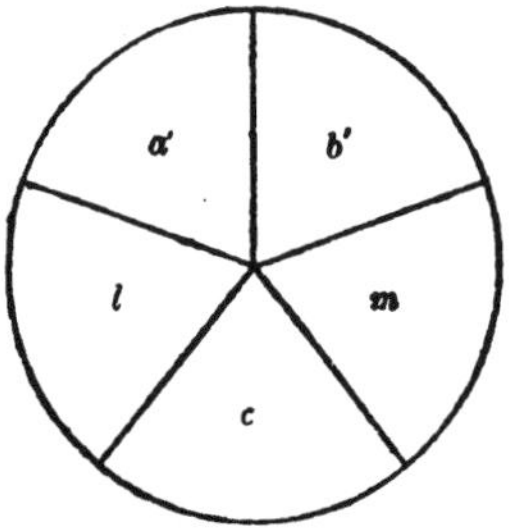


FIG. 39.

if we express all the formulae in terms of the segments $a'$, $l$, $c$, $m$, $b'$. The formulae become:

$$\cosh c = \sinh l \sinh m = \coth a' \coth b',$$

[1] John Napier, *Mirifici logarithmorum canonis descriptio*, Edinburgh, 1614 (lib. ii. cap. iv.). These rules have often been treated disparagingly by those (*e.g.* Airy and De Morgan) who saw in them only artificial mnemonics or at best curiosities with no fundamental scientific basis. The properties of the beautiful star-pentagon ("pentagramma mirificum"), by which these rules were originally established by Napier, were extensively studied by Gauss (*Werke*, iii. 481). The foundation of the rules for hyperbolic geometry was laid by Lobachevsky, *New Foundations of Geometry*, chap. x. He makes use of the diagrams in the first part of § 35. The modified forms of Napier's Rules were established by Engel in his edition of Lobachevsky's *New Foundations*, p. 345.

and four other pairs obtained by the cyclic permutation ($lma'cb'$) or ($a'lcmb'$).

If we write the five parts $a'$, $l$, $c$, $m$, $b'$ in cyclic order (Fig. 39), then

(B)
cosh (middle part) = product of hyp. sines of adjacent parts
= product of hyp. cotangents of opposite parts.

It is easily verified that this rule holds, with circular functions instead of hyperbolic functions, for a spherical triangle in euclidean space with hypotenuse $c$, sides $l'$ and $m'$ and opposite angles $a'$ and $b'$.

**34.** Expressing the formulae in terms of $\alpha'$, $\lambda$, $\gamma$, $\mu$, $\beta'$, we get, since

$$\cosh x = \operatorname{cosec} \xi, \quad \sinh x = \cot \xi, \quad \coth x = \sec \xi,$$

where $x$ stands for any one of the letters $a$, $b$, $c$, $l$, $m$, and $\xi$ for the corresponding Greek letter $\alpha$, $\beta$, $\gamma$, $\lambda$, $\mu$,

$$\sin \gamma = \tan \lambda \tan \mu = \cos \alpha' \cos \beta',$$

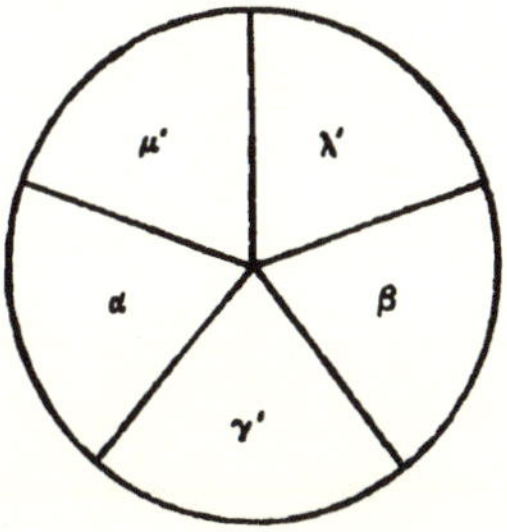


FIG. 40.

with four other pairs obtained by the cyclic permutation ($\alpha'\lambda\gamma\mu\beta'$). They may be read off from Fig. 40 by applying Rule (B) with circular functions.

But these are the formulae for a right-angled spherical triangle with hypotenuse $\gamma'$, sides $\alpha'$ and $\beta'$ and opposite angles $\mu'$ and $\lambda'$; or one with hypotenuse $\alpha$, sides $\mu$ and $\gamma$

and opposite angles $\lambda'$ and $\beta$. But this is just the spherical triangle which we found to correspond to the rectilinear triangle (§ 30). Hence *the formulae for a spherical triangle in hyperbolic space are exactly the same as those for a spherical triangle in euclidean space*, when we take as the measure of a side the angle which it subtends at the centre, and as the measure of an angle the dihedral angle between the planes passing through the sides and the centre. (Cf. Chap. III. § 21.) It may be noted that the letters $\mu'$, $a$, $\gamma'$, $\beta$, $\lambda'$ in Fig. 40 are the same, and in the same order, as those on the sides of the simple pentagon in Fig. 36.

## 35. Correspondence between a right-angled triangle and a tri-rectangular quadrilateral.

Draw $B\Omega \parallel CA$, and $D\Omega \perp BA$ and $\parallel CA$ (Fig. 41). Then $AD = l$, $CB\Omega = \alpha$, and

$$\alpha - \mu = \Pi(c + l). \qquad (1)$$

Similarly $$\beta - \lambda = \Pi(c + m). \qquad (1')$$

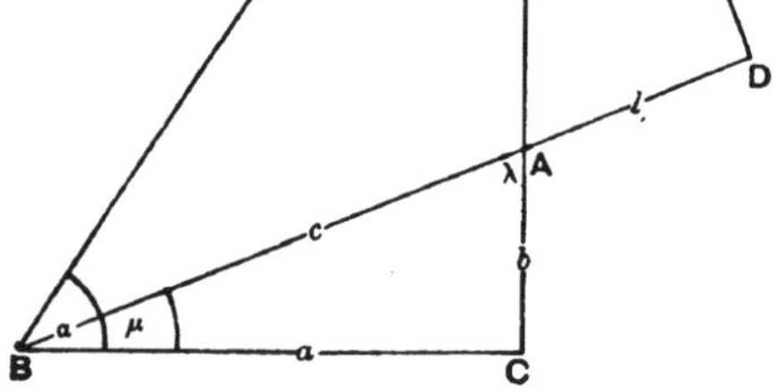


FIG. 41.

Draw $B\Omega \parallel AC$, and $D\Omega \perp BA$ and $\parallel AC$ (Fig. 42). Then $AD = l$, $CB\Omega = a$, and

$$a + \mu = \Pi(c - l). \quad \text{.......................(2)}$$

Similarly

$$\beta + \lambda = \Pi(c - m). \quad \text{.......................(2')}$$

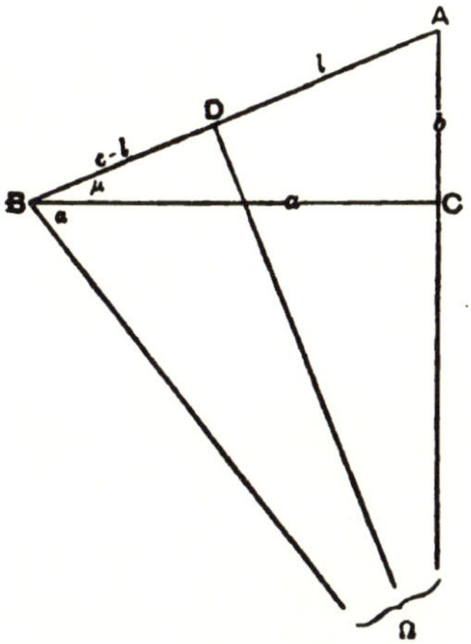


FIG. 42.

*Note.*—If $l > c$, then $\pi - (a + \mu) = \Pi(c - l)$;

if $l = c$, $a + \mu = \Pi(0) = \frac{\pi}{2}$;

which are both contained in (2) if we understand that

$$\Pi(-x) = \pi - \Pi(x).$$

Draw $D\Omega \perp CA$ and $\parallel BA$, and $E\Omega \perp BC$ and $\parallel BA$ (Fig. 43).

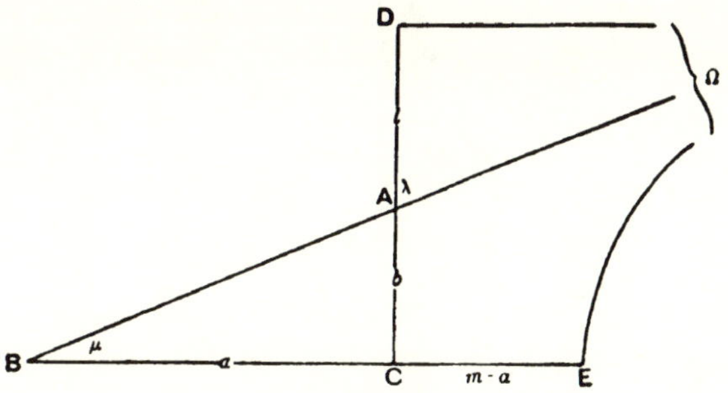


FIG. 43.

Then $AD=l$, $BE=m$, $EC\Omega=\Pi(m-a)$, $DC\Omega=\Pi(l+b)$, and

$$\Pi(m-a)+\Pi(l+b)=\frac{\pi}{2}. \quad \text{.....................(3)}$$

Similarly $$\Pi(m+a)+\Pi(l-b)=\frac{\pi}{2}. \quad \text{.....................(3')}$$

In the tri-rectangular quadrilateral with angle $\theta$ and sides in order

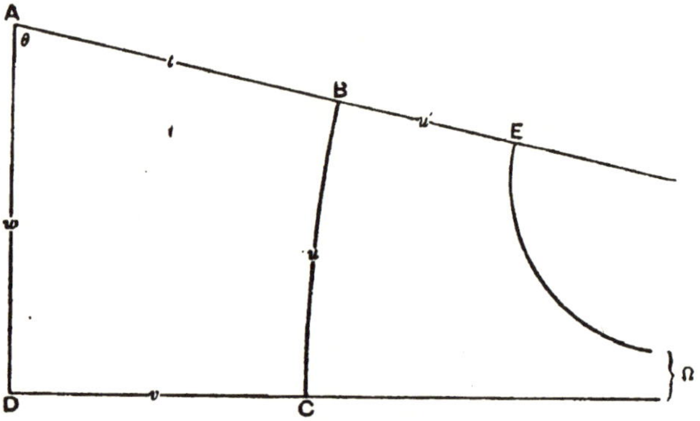


FIG. 44.

$t$, $u$, $v$, $w$, draw $E\Omega \parallel DC$ and $\perp AB$ (Fig. 44). Then $BE=u'$, for $\Pi(u)+\Pi(u')=\frac{\pi}{2}$, $DA\Omega=\Pi(w)$, $BA\Omega=\Pi(t+u')$, and

$$\Pi(w)+\Pi(t+u')=\theta. \quad \text{........................(I)}$$

Similarly $$\Pi(t)+\Pi(w+v')=\theta. \quad \text{........................(II)}$$

Draw $A\Omega \parallel CD$, and $E\Omega \perp AB$ and $\parallel CD$ (Fig. 45). Then $BE=u'$, $DA\Omega=\Pi(w)$, $BA\Omega=\Pi(t-u')$, and

$$\Pi(w)+\theta=\Pi(t-u'). \quad \text{.....................(I')}$$

Similarly $$\Pi(t)+\theta=\Pi(w-v'). \quad \text{.....................(II')}$$

Draw $E\Omega \perp CD$ and $\parallel BA$, $F\Omega \perp DA$ and $\parallel BA$ (Fig. 46). Then $EC=u'$, and if $AF=f$, $\theta=\Pi(f)$, and we have

$$\Pi(u'-v)+\Pi(w+f)=\frac{\pi}{2}. \quad \text{..................(III)}$$

Similarly $$\Pi(v'-u)+\Pi(t+f)=\frac{\pi}{2}. \quad \text{..................(III')}$$

Now the quadrilateral is determined by $t$ and $u$. Let $t=c$ and $u=m'$. Then

$$\Pi(c+m)=\theta-\Pi(w)=\beta-\lambda, \text{ from (I) and (1').}$$
$$\Pi(c-m)=\theta+\Pi(w)=\beta+\lambda, \text{ from (I') and (2').}$$

Therefore $\theta=\beta$ and $\Pi(w)=\lambda$ or $w=l$ and $f=b$.

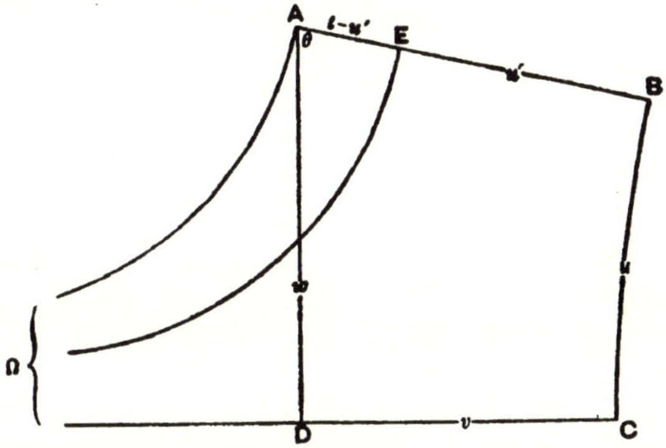


FIG. 45.

Then, comparing (III) and (3), we have

$$\Pi(m-v)=\frac{\pi}{2}-\Pi(l+b)=\Pi(m-a)\text{; therefore } v=a.$$

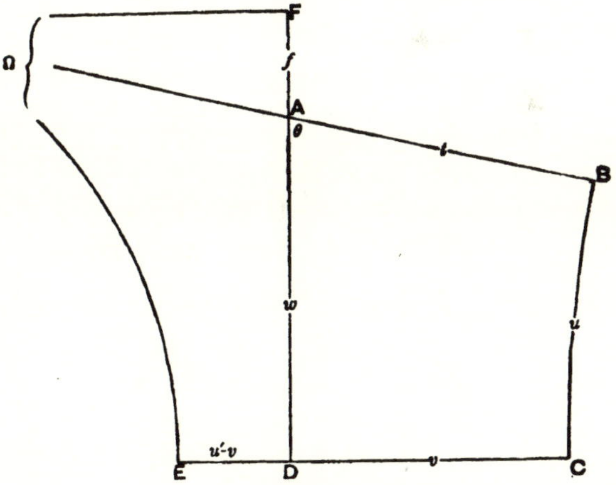


FIG. 46.

Hence, to a right-angled triangle ($c$, $a\ \lambda$, $b\ \mu$) there corresponds a tri-rectangular quadrilateral with angle $\beta$ and sides, starting from the angle, $c$, $m'$, $a$, $l$. By reversing the order of the sides, we get the quadrilateral ($\beta$, $l\ a\ m'\ c$), to which corresponds a triangle ($l$, $m'\gamma$, $b\ \alpha'$), or ($l$, $b\ \alpha'$, $m'\gamma$), Fig. 47.

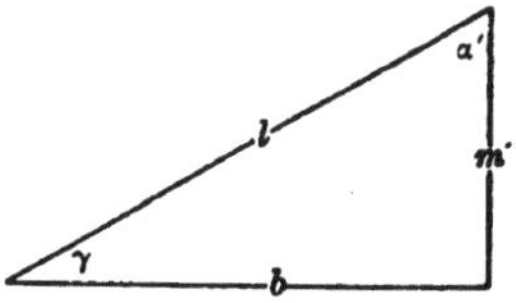


FIG. 47.

If we take $a'$, $l$, $c$, $m$, $b'$ as quantities determining the parts $a$, $\lambda$, $c$, $\mu$, $b$ of the triangle, then we get a triangle corresponding to the quantities $b'$, $a'$, $l$, $c$, $m$, and similarly, by cyclic permutation, we get five associated triangles. This forms an independent proof of the result deduced in § 31 from spherical triangles.

**36.** We can deduce from this correspondence that the relations between the parts of a tri-rectangular quadrilateral can be written down by rules exactly analogous to Napier's rules. If the angle is $C$ and the sides in order are $a$, $m$, $l$, $b$, write down in a circle the parts $\overline{C}$, $a$, $\overline{m}$, $\overline{l}$, $b$. Then

sine (middle part) = product of cosines of opposite parts

= product of tangents of adjacent parts,

with the same understanding as in the case of the triangle (§ 33 (A)).

If we write the parts in the cyclic order $C$, $l$, $\bar{a}$, $\bar{b}$, $m$, we get rules analogous to (B) at the end of § 33, viz.:

cos (middle part) = product of sines of adjacent parts

= product of cotangents of opposite parts.

**37.** The formulae for a general triangle can be obtained from those for a right-angled triangle by dividing the triangle into two right-angled triangles (Fig. 48).

Thus, $\sinh p = \sinh a \sin B = \sinh b \sin A$.

Hence $$\frac{\sinh a}{\sin A} = \frac{\sinh b}{\sin B} = \frac{\sinh c}{\sin C}.$$

Again, $\cosh c = \cosh c_1 \cosh c_2 + \sinh c_1 \sinh c_2,$

$$\cos C = \cos C_1 \cos C_2 - \sin C_1 \sin C_2.$$

Also $\cosh a = \cosh c_1 \cosh p, \quad \cosh b = \cosh c_2 \cosh p,$

$$\sinh c_1 = \sinh a \sin C_1, \quad \sinh c_2 = \sinh b \sin C_2,$$

$$\cos C_1 = \coth a \tanh p, \quad \cos C_2 = \coth b \tanh p.$$

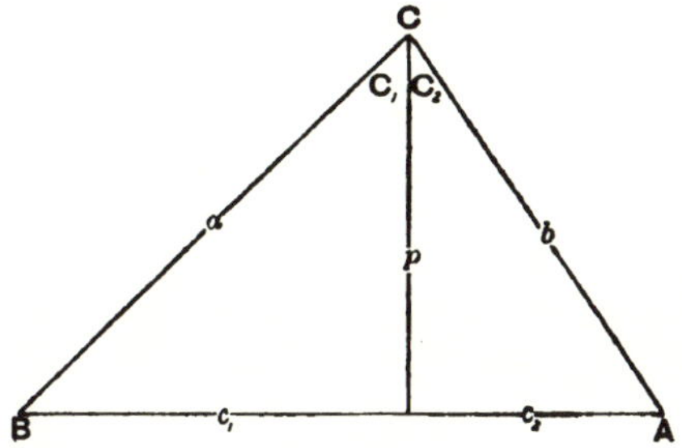


FIG. 48.

Therefore $\cosh c = \cosh a \cosh b \operatorname{sech}^2 p$

$$+ \sinh a \sinh b \sin C_1 \sin C_2$$

$$= \cosh a \cosh b \operatorname{sech}^2 p + \sinh a \sinh b$$

$$\times (\coth a \coth b \tanh^2 p - \cos C)$$

$$= \cosh a \cosh b - \sinh a \sinh b \cos C.$$

Similarly

$$\cos C = \cos A \cos B - \sin A \sin B \cosh c.$$

It is needless to write down other formulae, which may be obtained from the corresponding formulae of spherical trigonometry by putting cosh for cos and $i$ sinh for sin, when operating upon the sides, leaving the functions of the angles unaltered.

**38.** The formulae of hyperbolic trigonometry become those of euclidean plane trigonometry when the constant $k \to \infty$.

To a first approximation

$$\sinh\frac{a}{k}=\frac{a}{k}, \quad \cosh\frac{a}{k}=1+\frac{1}{2}\frac{a^2}{k^2}.$$

The formula

$$\cosh\frac{c}{k}=\cosh\frac{a}{k}\cosh\frac{b}{k}-\sinh\frac{a}{k}\sinh\frac{b}{k}\cos C$$

becomes
$$1+\frac{1}{2}\frac{c^2}{k^2}=\left(1+\frac{1}{2}\frac{a^2}{k^2}\right)\left(1+\frac{1}{2}\frac{b^2}{k^2}\right)-\frac{a}{k}\cdot\frac{b}{k}\cos C,$$

or
$$c^2=a^2+b^2-2ab\cos C.$$

This shows that when we are dealing with a small region, *i.e.* small in comparison with $k$, the geometry is sensibly the same as that of Euclid.

## 39. Circumference of a circle.

Let $ds$ be the length of the arc $PQ$ of a circle of radius $r$, which subtends an angle $d\theta$ at the centre. Then

$$\sinh\frac{1}{2}\frac{ds}{k}=\sinh\frac{r}{k}\sin\tfrac{1}{2}d\theta,$$

or
$$ds=k\sinh\frac{r}{k}\,d\theta.$$

Hence the length of the whole circumference is $2\pi k\sinh\frac{r}{k}$.

Here, for the first time, we require to consider the actual value of $\pi$, for the formula $\lim\limits_{\theta\to 0}\frac{\sin\theta}{\theta}=1$, which is here assumed, is true only when the number of units in a flat-angle is the transcendental irrational number 3·14159....

Draw $PN\perp OA$ and $PT$ the tangent at $P$. Let the arc $AP=s$, $PN=y$ and $PT=t$. Then

$$s=k\theta\sinh\frac{r}{k},$$

$$\sinh\frac{y}{k}=\sinh\frac{r}{k}\sin\theta,$$

$$\sinh\frac{r}{k}=\tanh\frac{t}{k}\cot\theta.$$

Let the centre $O$ go to infinity, so that the circle becomes a horocycle. Then $r \to \infty$, $\theta \to 0$, and

$$k \sinh \frac{y}{k} = s \cdot \frac{\sin \theta}{\theta} \to s \quad \text{and} \quad k \tanh \frac{t}{k} = s \cdot \frac{\tan \theta}{\theta} \to s.$$

Comparing these with the formulae in § 28, we find $S = k$.

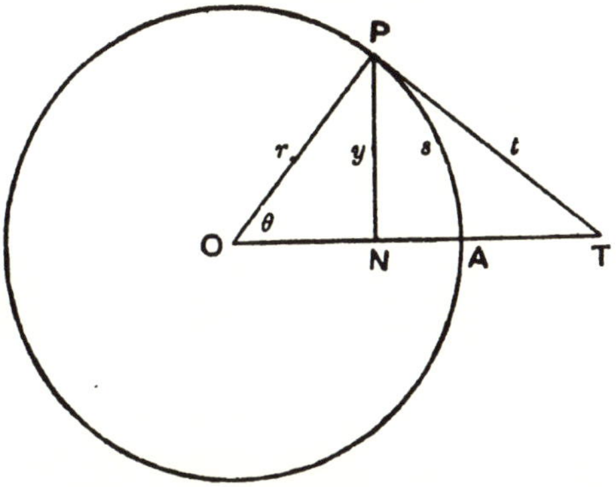


FIG. 49.

## 40. Sum of the angles and area of a triangle.

Join $MN$, the middle points of $AB$, $AC$, and construct the equidistant-curve with $MN$ as axis, which passes through

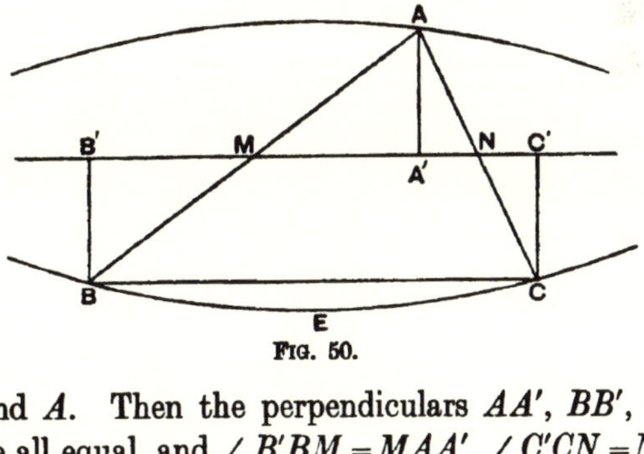


FIG. 50.

$B$, $C$ and $A$. Then the perpendiculars $AA'$, $BB'$, $CC'$ to $MN$ are all equal, and $\angle B'BM = MAA'$, $\angle C'CN = NAA'$.

Denote by $ABE$ the angle which $AB$ makes with the tangent to the equidistant-curve at $B$; the angle $B'BE$ is a right angle. Then

$$\angle BAC + ABE + ACE = B'BM + MBE + C'CN + NCE = \pi.$$

Hence the sum of the angles of the triangle $ABC$ $= \pi - 2CBE$. The difference $\pi - (A + B + C)$ is called the *defect* of the triangle.

Again, the area of the triangle $ABC$

$$= BMNC + MAA' + NAA' = B'BCC'.$$

Hence all triangles with base $BC$ and vertex on the other branch of the equidistant-curve which passes through $B$, $C$ and $A$ have the same area and the same angle-sum or defect.

Now, if we are given any two triangles, we can transform one of them into another of the same area and defect, and having one of its sides equal to one of the sides of the other triangle.

Let $ABC$, $DEF$ be the two triangles, and let $DF$ be the greatest of the six sides. Construct an equidistant-curve passing through $B$, $C$ and $A$. With centre $C$ and radius equal to $DF$, draw a circle cutting the branch of the equidistant-curve on which $A$ lies in $A'$. Then the triangle $A'BC$ has the same area and defect as the triangle $ABC$, and has the side $A'C$ equal to $DF$.

Again, if the perpendicular bisector of the base $BC$ of a triangle $ABC$ meets the other branch of the equidistant-curve in $A'$, the isosceles triangle $A'BC$ has the same area and defect as the triangle $ABC$.

Hence, if two triangles have the same area they can be transformed into the same isosceles triangle, and have therefore the same defect, and conversely.

Now, let a triangle $ABC$ with area $\Delta$ and defect $\delta$ be divided into two triangles $ABD$, $ADC$ with areas $\Delta_1$ and $\Delta_2$ and defects $\delta_1$ and $\delta_2$.

Then
$$\delta_1 = \pi - BAD - B - ADB,$$
$$\delta_2 = \pi - DAC - C - ADC.$$

Therefore $\delta_1 + \delta_2 = 2\pi - A - B - C - \pi = \pi - A - B - C = \delta,$

and
$$\Delta_1 + \Delta_2 = \Delta.$$

If $\Delta_1 = \Delta_2$, then $\delta_1 = \delta_2$ and $\Delta = 2\Delta_1$, $\delta = 2\delta_1$.

Hence the defect is proportional to the area, or

$$\Delta = \lambda(\pi - A - B - C).$$

The value of this constant $\lambda$ depends upon the units of angle and area which are employed; but when these have been chosen it is given absolutely.

**41. Relation between the units of length and area.**

In euclidean geometry the units of length and area are immediately connected by taking as the unit of area the area of a square whose side is the unit of length. In fact the relationship is so obvious that there is constant confusion, though we are not always aware of it, between the area of a rectangle and the product of two numbers. Thus modern treatment has tended to confuse the theorems of the second book of Euclid, which are purely geometrical theorems relating to areas of squares and rectangles, with algebraic theorems relating to "squares" and products of numbers. The expression "product of two lines" has no meaning until we frame a suitable definition consistent with the rest of the subject-matter. The area of a rectangle is not equal to the product of its sides, but the number of units of area in the area of a rectangle is equal to the product of the numbers of units of length in its sides.

It would take us too far out of our way to examine completely the notion of area. We shall simply take advantage of the fact, that when we are dealing with a very small region of the plane we can apply euclidean geometry. Thus, while there exists no such thing as a euclidean square in non-euclidean geometry, if we take a regular quadrilateral[1] with all its sides very small we may take as its area the square of the number of units of length in its sides; or, more accurately, the units of length and area are so adjusted that the ratio of the area of a regular quadrilateral to the square of the number of units of length in its side tends to the limit unity as the sides are indefinitely diminished.

Let us apply this to find the area of a sector of a circle $POQ$, the angle $POQ = \theta$ being very small.

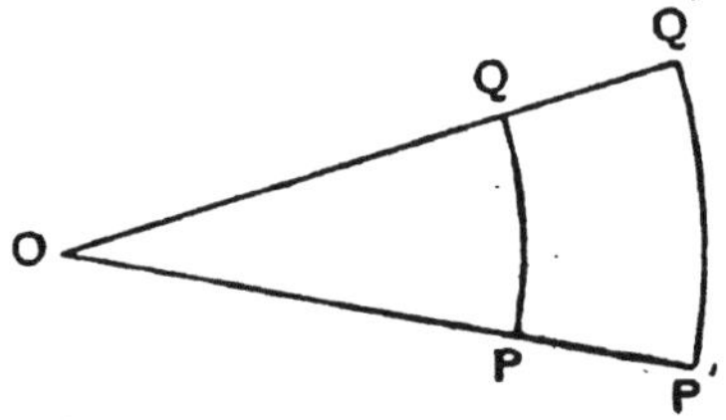

FIG. 51.

Produce $OP$, $OQ$ to $P'$ and $Q'$. Let $OP = OQ = r$, $PP' = QQ' = dr$ Then

$$\text{area of } PQQ'P' = dr \, . \, PQ = k\theta \, dr \sinh \frac{r}{k}.$$

$$\text{Hence the area of the sector} = k^2\theta\left(\cosh \frac{r}{k} - 1\right)$$

$$= 2k^2\theta \sinh^2 \frac{r}{2k},$$

and the area of the whole circle is $4\pi k^2 \sinh^2 \dfrac{r}{2k}$.

[1] A regular polygon is one which has all its sides equal and all its angles equal.

We can apply this now to find the area of a triangle by another method. It is sufficient to take a triangle $ABC$ with a right angle at $C$. Divide it into small sectors by lines drawn through $A$. Then the area is given by

$$\int_0^A k^2 dA\left(\cosh\frac{c}{k}-1\right).$$

Express $c$ in terms of $A$ and the constant $b$, write $\tanh\frac{b}{k}=t$, and put $\cos^2 A=y$, and we get, after some reductions,

$$\cosh\frac{c}{k}\,dA=\frac{-dy}{2\sqrt{1-y}\sqrt{y-t^2}}=\frac{-dy}{2\sqrt{\frac{1}{4}(1-t^2)^2-(y-\frac{1}{2}\overline{1+t^2})^2}}.$$

The integral of this term, from $y=1$ to $y=\cos^2 A$, is

$$\frac{1}{2}\cos^{-1}\frac{2y-(1+t^2)}{1-t^2}=\frac{1}{2}\cos^{-1}\left(1-2\cosh^2\frac{b}{k}\sin^2 A\right)$$

$$=\tfrac{1}{2}\cos^{-1}(1-2\cos^2 B)=\tfrac{1}{2}(\pi-2B).$$

Hence the area of the triangle

$$=k^2\left(\frac{\pi}{2}-B-A\right)=k^2\left(\pi-\frac{\pi}{2}-A-B\right).$$

**42.** It appears then that, when the angles are measured in "circular" measure, the constant $\lambda=k^2$, and the formula for the area of a triangle becomes

$$\Delta=k^2(\pi-A-B-C).$$

As the area of a triangle increases the sum of the angles diminishes, but, so long as the vertices are real, the angles are positive quantities; the area cannot therefore exceed $\pi k^2$. This is therefore the maximum limit to the area of a triangle when its angles all tend to zero. A triangle of maximum area has all its vertices at infinity and its sides are parallel in pairs.

**43**. On account of its neatness, we add the proof that Gauss gave of the formula for the area of a triangle, in a letter[1] to W. Bolyai acknowledging the receipt of the "Appendix."

Gauss starts by assuming that the area enclosed by a straight line and two lines through a point parallel to it is finite, and a certain function $f(\pi-\phi)$ of the angle $\phi$ between the two parallels; and further, that the area of a triangle whose vertices are all at infinity is a certain finite quantity $t$

Then we have, from Fig. 52, $f(\pi-\phi)+f(\phi)=t$.

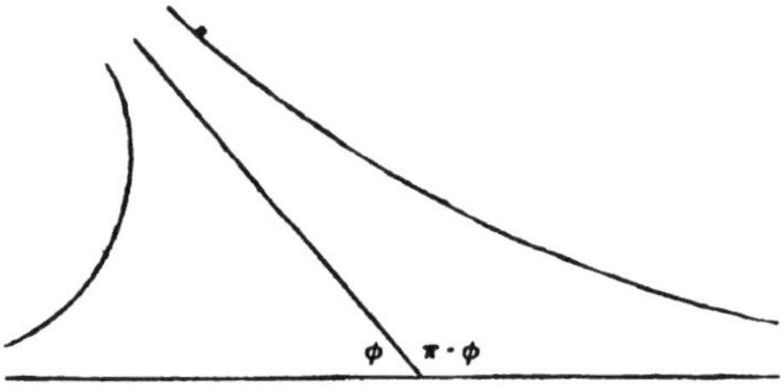


FIG. 52.

Again, from Fig. 53, $f(\phi)+f(\psi)+f(\pi-\phi-\psi)=t$.

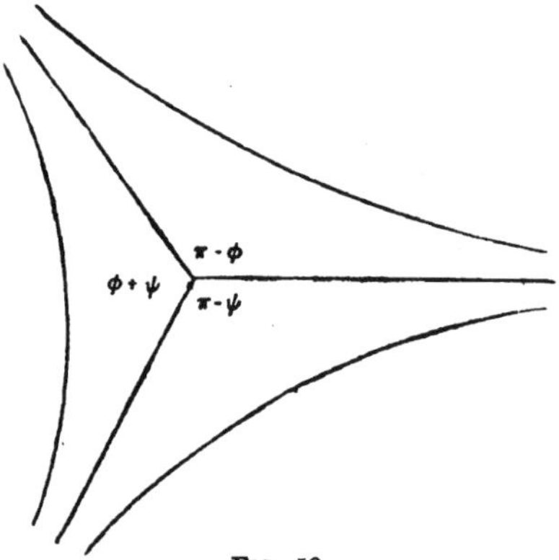


FIG. 53.

Hence $$f(\phi)+f(\psi)=f(\phi+\psi).$$

Whence $$f(\phi)=\lambda\phi,$$

where $\lambda$ is a constant, and therefore $t=\lambda\pi$.

[1] 6th March, 1832 (Gauss' *Werke*, viii. 221).

Now, by producing the sides of any triangle with angles $\alpha$, $\beta$, $\gamma$, and drawing parallels, we have (Fig. 54)

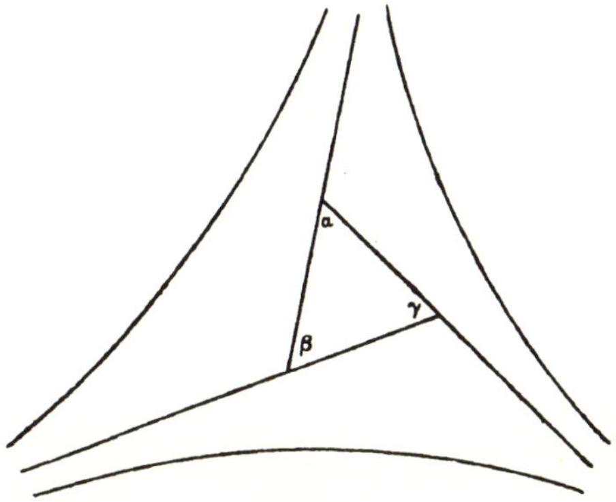


FIG. 54.

$$t = f(\alpha) + f(\beta) + f(\gamma) + \Delta.$$

Therefore $$\Delta = \lambda(\pi - \alpha - \beta - \gamma).$$

**44. Area of a polygon.**

The area of a polygon can be found by breaking it up into triangles. By joining one vertex to each of the others, we divide an $n$-gon into $n-2$ triangles. The sum of the angles of the $n$-gon is equal to the sum of the angles of the $n-2$ triangles.

Let $\Delta_1$, $\Delta_2$, ... be the areas, and $\delta_1$, $\delta_2$, ... the defects of these triangles; then, if $S$ is the sum of the angles and $A$ the area of the polygon,

$$A = \Sigma\Delta = \Sigma k^2\delta = k^2(\overline{n-2}\,.\,\pi - S).$$

If $S'$ is the sum of the exterior angles, $S' + S = n\pi$; therefore

$$A = k^2(S' - 2\pi),$$

which is independent of $n$.

**45**. We add here another proof of the result that *the geometry of horocycles on the horosphere is the same as the geometry of straight lines on the euclidean plane.*

Let the three parallel lines in space $A\Omega$, $B\Omega$, $C\Omega$ be cut by a

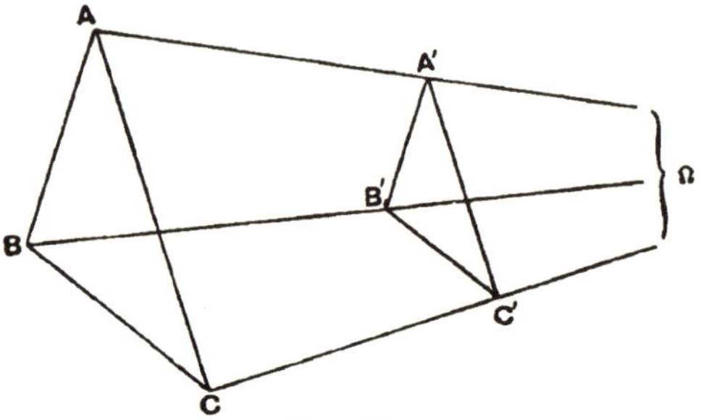


FIG. 55.

horosphere with centre $\Omega$ in $A$, $B$, $C$, and make $AA'=BB'=CC'$, so that $A'B'C'$ lie again on a horosphere with centre $\Omega$. (See Ex. 8.)

Let the dihedral angles between the planes $BC\Omega$, $CA\Omega$, $AB\Omega$ be $\alpha$, $\beta$, $\gamma$, and let the angles of the rectilinear triangle $A'B'C'$ be $\alpha'$, $\beta'$, $\gamma'$, and its area $\Delta$.

Then, as $AA'$ increases, the angles $\Omega A'B'$, $\Omega A'C'$, etc., all tend to right angles; hence $\alpha'$, $\beta'$, $\gamma'$ tend to the values $\alpha$, $\beta$, $\gamma$. Also $\Delta \to 0$.

Now $\Delta = k^2(\pi - \alpha' - \beta' - \gamma')$; hence $\alpha + \beta + \gamma = \pi$, *i.e.* when three planes intersect in pairs in three parallel lines the sum of the dihedral angles is equal to two right angles. Hence the sum of the angles of a geodesic triangle on the horosphere is equal to two right angles.

## EXAMPLES II.

**1.** Prove that the four axes of the circumcircles of a triangle form a complete quadrilateral whose diagonal triangle is the given triangle; and state the reciprocal theorem.

**2.** If a simple quadrilateral is inscribed in a circle, horocycle or one branch of an equidistant-curve, prove that the sum of one pair of opposite angles is equal to the sum of the other pair of opposite angles. Show that this holds also for a crossed quadrilateral if the angles are measured always in the same sense, and for a quadrilateral whose vertices are distributed between the two branches of an equidistant-curve if the angles on opposite branches are reckoned of opposite sign.

**3.** If a simple quadrilateral is circumscribed about a circle, prove that the sum of one pair of opposite sides is equal to the sum of the other pair of opposite sides. Examine the case of a crossed quadrilateral circumscribed about a circle, equidistant-curve or horocycle.

**4.** If $a$ is the chord of an arc $\alpha$ of a horocycle, prove that

$$\alpha = 2k \sinh \tfrac{1}{2}a/k.$$

**5.** If $\theta$ is the angle which the chord of a horocycle makes with the tangent at either end, and $\alpha$ is the arc, prove that $\alpha = 2k \tan \theta$.

**6.** If $\phi$ is the angle which the tangent at one extremity of an arc $\alpha$ of a horocycle makes with the radius through the other extremity, prove that $\alpha = k \cos \theta$.

**7.** Prove that the arc of an equidistant-curve of distance $a$, corresponding to a segment $x$ on its axis, is $x \cosh a/k$.

**8.** If $A$, $B$ are corresponding points on the parallels $AA'$, $BB'$, and $A$, $C$ are corresponding points on the parallels $AA'$, $CC'$, prove that $B$, $C$ are corresponding points on the parallels $BB'$, $CC'$.

**9.** Prove the following construction for the parallel from $O$ to $NA$. Draw $ON \perp NA$. Take any point $A$ on $NA$, draw $OB \perp ON$ and $AB \perp OB$. With centre $O$ and radius equal to $NA$, draw a circle cutting $AB$ in $P$. Then $OP \parallel NA$.

**10.** Prove that the radius of the inscribed circle of a triangle of maximum area is $\tfrac{1}{2}k \log_e 3$.

**11.** In a quadrilateral of maximum area, if $2a$, $2b$ are the lengths of the common perpendiculars of opposite sides, prove that

$$\sinh \frac{a}{k} \sinh \frac{b}{k} = 1.$$

**12.** A regular quadrilateral is symmetrically inscribed in a regular maximum quadrilateral; prove that each of its angles is $\cos^{-1}\tfrac{1}{3}$.

**13.** If the three escribed circles of a triangle are all horocycles, prove that each side of the triangle is $\cosh^{-1}\tfrac{5}{4}$, and that the radius of the inscribed circle is $\tanh^{-1}\tfrac{1}{4}$, and the radius of the circumcircle is $\tanh^{-1}\tfrac{1}{2}$ ($k$ being unity).

**14.** In euclidean geometry prove that any convex quadrilateral can by repetition of itself be made to cover the whole plane without overlapping.

**15.** In hyperbolic geometry prove that any convex polygon with an even number of sides can by repetition of itself be made to cover the whole plane without overlapping, provided the sum of its angles is equal to or a submultiple of four right angles. Show that the same is possible if the number of sides is odd, provided the sum of the angles is equal to or a submultiple of two right angles.

**16.** If $a$ is the side and $\alpha$ the angle of a regular $n$-gon, prove that

$$\cos\frac{\pi}{n} = \sin\frac{\alpha}{2}\cosh\frac{a}{2k}.$$

**17.** If $r$ is the radius of the inscribed circle, $R$ that of the circumscribed circle of a regular $n$-gon with side $a$ and angle $\alpha$, prove that

$$\sinh\frac{r}{k} = \cot\frac{\pi}{n}\tanh\frac{a}{2k}, \quad\text{and}\quad \cosh\frac{R}{k} = \cot\frac{\pi}{n}\cot\frac{\alpha}{2}.$$

**18.** A regular network is formed of regular $n$-gons, $p$ at each point. Show that the area of each polygon is $k^2\pi(2n/p - n + 2)$.

**19.** A semiregular network is formed of triangles and hexagons with the same length of side, three of each being at each point. Prove that the length of the side is $2k\cosh^{-1}\sqrt{\tfrac{1}{3}(4+\sqrt{3})}$.

**20.** A semiregular network consists of regular polygons all with the same length of side. At each vertex there are $p_1$ $n_1$-gons, $p_2$ $n_2$-gons, $p_3$ $n_3$-gons, etc. If each $n_i$-gon has area $A_i$, prove that

$$\Sigma A\frac{p}{n} = \pi k^2\left[2\left(1+\Sigma\frac{p}{n}\right) - \Sigma p\right].$$

**21.** If a ring of $n$ equal circles can be placed round an equal circle, each one touching the central circle and two adjacent ones, prove that the radius of each circle is given by $2\cosh\frac{r}{k}\sin\frac{\pi}{n} = 1$.

**22.** Prove that the area included between an arc of an equidistant-curve of distance $a$, its axis, and two ordinates at distance $x$ is

$$kx\sinh\frac{a}{k}.$$

**23.** $AA' \parallel BB'$ and they make equal angles with $AB$. $AC \perp BB'$ and $AD \perp AC$. If the angle $A'AD = z$, prove that the area of the circle whose radius is $AB$ is equal to $\pi k^2 \tan^2 z$. (J. Bolyai.)

**24.** Prove that the volume of a sphere of radius $r$ is

$$\pi k^3 \left(\sinh \frac{2r}{k} - \frac{r}{k}\right).$$

**25.** These parallel lines $A\Omega$, $B\Omega$, $C\Omega$ are cut by two horocycles with centre $\Omega$ in $A$, $B$, $C$ and $A'$, $B'$, $C'$. Prove that the arcs $AB : BC = A'B'$ $B'C'$.

**26.** $A_1B_1$, $A_2B_2$, $A_3B_3$ are arcs of concentric horocycles as in Fig. 29, and $A_1A_2 = A_2A_3$. Prove that $A_1B_1 : A_2B_2 = A_2B_2 : A_3B_3$. Hence show that the ratio $A_1B_1 : A_2B_2$ depends only on the length of $A_1A_2$.

**27.** Prove that the sides of a pentagon whose angles are all right angles are connected by the relations

cosh (middle side) = product of hyp. cots. of adjacent sides
= product of hyp. sines of opposite sides.

**28.** A simple spherical pentagon, each of whose vertices is the pole of the opposite side, is projected from the centre upon any plane. Prove that the projection is a pentagon whose altitudes are concurrent; and that the product of the hyp. tangents of the segments into which each altitude is divided is the same.

# CHAPTER III.

## ELLIPTIC GEOMETRY.

1. The hypothesis of elliptic geometry is that the straight line, instead of being of infinite length, is closed and of finite length. Two straight lines in the same plane will always meet, even when they are both perpendicular to a third straight line.

Let $l$, $m$, $n$, be three straight lines drawn perpendicular to

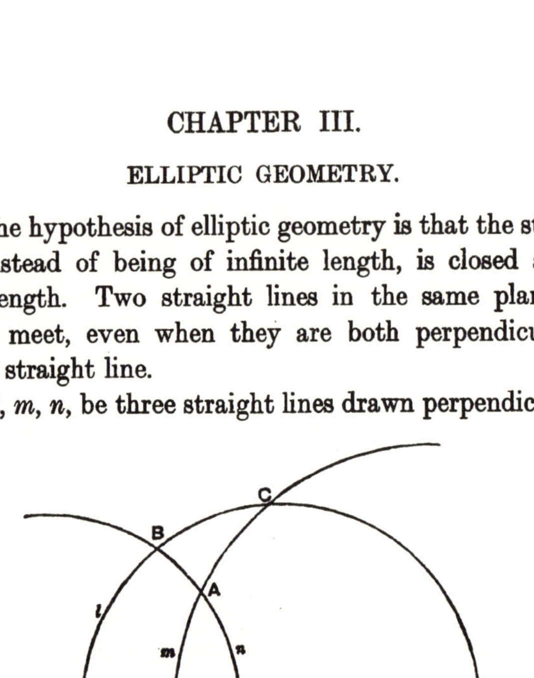


FIG. 56.

another straight line $a$ at the points $L$, $M$, $N$. Let $m$, $n$ meet in $A$; $n$, $l$ in $B$; and $l$, $m$ in $C$.

When $LB$ is produced it will meet $a$ again either in $L$ or in some other point. Let $L'$ be the first point in which it again meets $a$.

Then, from isosceles triangles, we have $BL = BN = BL'$, $CL = CM = CL'$. Hence $B$ and $C$ are both the middle point of the segment $LBL'$, and must therefore coincide. In the same way $A$, $B$ and $C$ all coincide.

Hence all the perpendiculars to a given line $a$ on one side of it meet in a point $A$, and $A$ is equidistant from all points on the line. The point $A$ is called the *absolute pole* of the line $a$, and $a$ is called the *absolute polar* of $A$. If $P$ is any point on $a$, the distance $AP$ is called a *quadrant*, and $A$ is said to be *orthogonal* to $P$, or $A$ and $P$ are called *absolute conjugate points.*

**2.** The perpendiculars drawn in the other sense will similarly meet in a point $A'$.

The question arises : are $A$ and $A'$ distinct points ?

On the hypothesis that $A$ and $A'$ are distinct points, two straight lines have two points in common. It could be proved that in this case any two straight lines would intersect in a pair of points distant from one another two quadrants. A consistent system of geometry results, which is exactly like the geometry on a sphere, straight lines being represented by great circles, and is therefore called SPHERICAL GEOMETRY. The two points of intersection of two lines are called *antipodal points.* Two points determine a line uniquely except when they are antipodal points ; a pair of antipodal points determine a whole pencil of lines.

On the hypothesis that $A$ and $A'$ are one and the same point, two straight lines always cut in just one point, and two distinct points uniquely determine a line. This gives again a consistent system of geometry, which is called ELLIPTIC GEOMETRY.[1]

[1] Sometimes both of these systems are called Elliptic geometry, and they are distinguished as the Antipodal or Double form and the Polar or Single form. We shall, however, keep the term Elliptic geometry for the latter form.

While spherical geometry admits more readily of being realised by means of the sphere, elliptic geometry is by far the more symmetrical, and our attention will be confined entirely to this type. Elliptic geometry has also the advantage that it more nearly resembles euclidean geometry, since in euclidean geometry all the perpendiculars to a straight line in a plane have to be regarded as passing through one point (at infinity).

Another mode of representation of these two geometries exists, which exhibits them both with equal clearness. Consider a bundle of straight lines and planes through a point $O$. If we call a straight line of the bundle a "point," and a plane of the bundle a "line," we have the following theorems with their translations. (Cf. Chap. II. § 24.)

| | |
|---|---|
| Two lines through $O$ uniquely determine a plane through $O$. | Two "points" uniquely determine a "line." |
| Two planes through $O$ intersect always in a single line through $O$. | Two "lines" intersect always in a single "point." |
| All the planes through $O$ perpendicular to a given plane $a$ through $O$ pass through a fixed line $a$ through $O$, which is orthogonal to every line through $O$ lying in $a$. | All the "lines" perpendicular to a given "line" $a$ pass through a fixed "point" $A$, which is orthogonal to every "point" lying in $a$. |

Hence elliptic geometry can be represented by the geometry of a bundle of lines and planes. In the same way spherical geometry can be represented by the geometry of a bundle of rays (or half-lines) and half-planes. Two rays which together form one and the same straight line represent a pair of antipodal points.

In elliptic geometry all straight lines are of the same finite length $2q$, equal to two quadrants.

If we extend these considerations to three dimensions, all the perpendiculars to a plane $\alpha$ pass through a point $A$, the absolute pole of $\alpha$, and the locus of points a quadrant distant from a point $A$ is a plane $\alpha$, the absolute polar of $A$.

3. The plane in elliptic geometry, or, as we may call it, the elliptic plane, differs in an important particular from the euclidean or hyperbolic plane. It is not divided by a straight line into two distinct regions.

Imagine a set of three rectangular lines $Oxyz$ with $Oy$ on the line $AM$ and $Oz$ always cutting the fixed line $AP$.

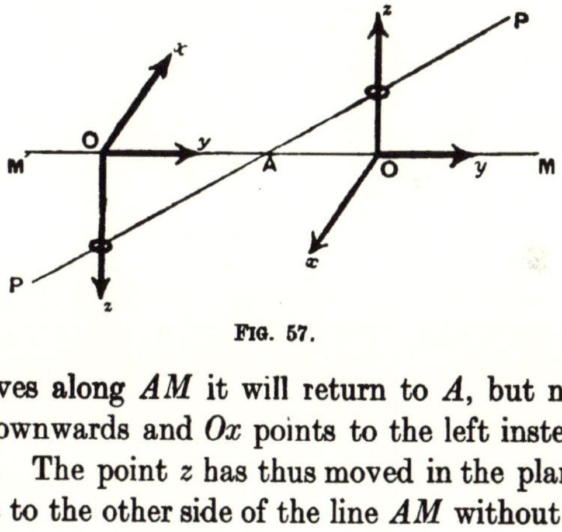


FIG. 57.

As $O$ moves along $AM$ it will return to $A$, but now $Oz$ is turned downwards and $Ox$ points to the left instead of to the right. The point $z$ has thus moved in the plane $PAM$ and come to the other side of the line $AM$ without actually crossing it.

A concrete illustration of this peculiarity is afforded by what is called Möbius' sheet, which consists of a band of paper half twisted and with its ends joined. A line traced along the centre of the band will return to its starting point, but on the opposite surface of the sheet. The two

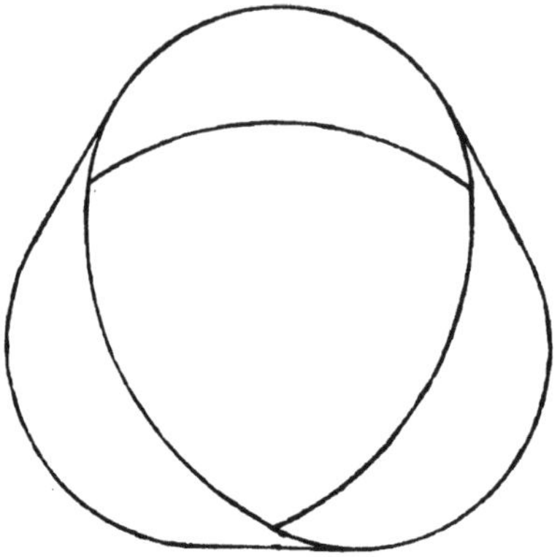

FIG. 58.

sides of the sheet are continuously connected. The elliptic plane is therefore a one-sided surface.

If we carry out the same procedure for the euclidean plane, we shall obtain exactly similar results, with the exception that a point passes through infinity in going from one side of the line to the other. This is well illustrated by the case of a curve which runs along an asymptote. Ordinarily the curve lies on opposite sides of the asymptote at the two ends, and thus appears to cross the asymptote. When it does actually cross the asymptote at infinity it has a point of inflexion there and lies on the same side of the asymptote at each end.

### 4. Absolute polar system.

To every point in space corresponds a plane, and *vice versa*, which are absolute pole and polar.

If the polar of a point $A$ passes through $B$, the polar of $B$ will pass through $A$, because the distance $AB$ is a quadrant.

Let $A$, $B$ be two points on a line $l$; the polars of $A$ and $B$ intersect in a line $l'$. Let $A'$ and $B'$ be any two points on $l'$; then the polar of $A'$ passes through both $A$ and $B$. Hence the polars of all points on the line $l'$ pass through the line $l$. If $P$ is any point on $l$, its polar will pass through $A'$ and $B'$. Therefore the polars of all points on the line $l$ pass through the line $l'$.

To every line $l$, therefore, there corresopnds a line $l'$, the absolute polar of $l$, such that the polar of any point on $l$ passes through $l'$ and *vice versa*. All points of $l'$ are a quadrant distant from all points of $l$, and every line which meets both $l$ and $l'$ cuts them at right angles.

If we confine ourselves to a plane, to every point in the plane corresponds a line in the plane and *vice versa*.

These relations are exactly the same as those that we get in ordinary geometry by taking poles and polars with regard to a conic in a plane, or a surface of the second degree in space. The points on the conic or quadric surface have the property that they lie on their polars; the polar is a tangent to the conic or quadric and the pole is the point of contact.

### 5. Projective geometry.

These relations of polarity with regard to a conic belong to *pure projective geometry*, and have nothing whatever to do with actual measurement, distances or angles. All the theorems of projective geometry can be at once transferred to non-euclidean geometry, for, so long as we are not dealing with actual metrical relations, non-euclidean geometry is in no way whatever distinguished from euclidean. Pure projective geometry takes no notice of points at infinity, for infinity here implies infinite distance, and is therefore irrelevant to the subject. It has therefore nothing to do with parallel lines.

Unfortunately most English text-books on projective geometry start by assuming euclidean metrical geometry. A harmonic range is defined in terms of the ratios of segments, and a conic is obtained as the "projection" of a circle. This treatment unnecessarily limits the generality of projective geometry, and attaches a quite unmerited importance to euclidean metric.

The use of analytical geometry might be thought to supply a means for a general treatment, for the algebraic relations between numbers which express the relations of projective geometry are just theorems of arithmetic, and these may be applied to any subject matter which can be subjected to numerical treatment, whether that subject matter is euclidean or non-euclidean geometry. But the difficulty in applying this procedure is that the subject matter must first be prepared for numerical treatment. This means either postponing the introduction of projective geometry until metrical geometry, with a system of coordinates, has been established,[1] which is just the fault we wish to avoid, or the establishment of a system of projective coordinates independent of distance. In either case we have to assume much more than is really necessary.

For convenience of reference we shall give a summary of the theorems of projective geometry which we shall require, assuming that proofs of these are available which do not involve metrical geometry. (Reference may be made to Reye, *Geometry of Position*, Part I., translated by Holgate, New York, 1898, or Veblen and Young, *Projective Geometry*, Vol. I., Boston, 1910.)

If two ranges of points are made to correspond in such a way that to every point $P$ on the one range corresponds uniquely a point $P'$ on the other, and *vice versa*, the ranges are said to be *homographic*.

Notation. $\{P\} \barwedge \{P'\}$.

The simplest way of obtaining a range which is homo-

[1] Cf. Chap. IV. § 21.

graphic with a given range is as follows. Take any point $O$, not on the axis of the range; join $O$ to the points of the range, and cut these rays by any transversal. The range on this transversal is called the *projection* of the first range and is homographic with it. In this special position, in which the lines joining pairs of corresponding points are concurrent, the ranges are said to be in *perspective*, with centre $O$.

Notation. $\{P\} \overline{\wedge}_O \{P'\}$ or $\{P\} \overline{\wedge} \{P'\}$.

It can be proved that two homographic ranges can always be connected by a finite number of projections, and in fact this number can in general be reduced to two.

It can be proved that

$$(ABCD) \wedge (BADC) \wedge (CDAB) \wedge (DCBA),$$

but in general the four points are projective in no other order.

Properties which are unaltered by projection are called projective properties. Thus, points which are collinear, or lines which are concurrent, retain these properties after projection.

A *harmonic range* is projected into a harmonic range. We cannot define a harmonic range in terms of the ratios of segments, because a segment is not projective. We define a harmonic range thus: Let $X$, $Y$ be two given points on a line, and $P$ a third point. (See Fig. 85, Chap. IV.) Through $P$ draw any line $PST$, and on it take any two points $S$, $T$. Join $S$ and $T$ to $X$ and $Y$; let $SX$ cut $TY$ in $V$, and $SY$ cut $TX$ in $U$. Join $UV$, and let it cut $XY$ in $Q$. $Q$ is called the *harmonic conjugate* of $P$ with regard to $X$ and $Y$. This construction can be proved to be unique; $P$, $Q$ are distinct, and are separated by and separate $X$ and $Y$. If $(XY, PQ)$ is a harmonic range $(XY, PQ) \barwedge (XY, QP)$.

If we start with three points on a line, we can derive an indefinite number of other points by the above *quadrilateral construction*, and in fact we can find a new point lying between any two given points. All the points derived in this way form a *net of rationality*. They do not give all the points on the line. To secure this we would require an assumption of *continuity*.

If three points $A$, $B$, $C$ of one line are projected on to three points $A'$, $B'$, $C'$ of another line, the correspondence between all the points of the two ranges is determined. This is the *fundamental theorem* of projective geometry.

Two homographic ranges can exist on the same line. If three points $A$, $B$, $C$ are *self-corresponding*, it follows by the fundamental theorem that all the points are self-corresponding. Hence two homographic ranges on the same line cannot have more than *two* self-corresponding points.

That it is possible in certain cases to have two self-corresponding points is shown in Fig. 59. $l$ is the given line, $l_1$ an intermediate line on which a range of points $\{P\}$ is projected from centre $S_1$, and $S_2$ is a second centre of projection from which the projected range $\{P_1\}$ is projected on to $l$.

In this way $P'$ corresponds to $P$. Let $l_1$ cut $l$ in $Y$, and let $S_1S_2$ cut $l$ in $X$. Then $X$ and $Y$ are self-corresponding points. If $l_1$ passes through $X$, the two self-corresponding points will coincide.

If $\{P\}$ and $\{P'\}$ are two homographic ranges on the same line, such that to $P$ corresponds $P'$, in general to $P'$ will correspond another point $P''$. If $P''$ coincides with $P$, the points of the line are connected in pairs and are said to form an *involution*. If $D_1$ and $D_2$ are the double or self-corresponding points of an involution, and $X$, $X'$ are a pair of corresponding points, $(D_1D_2XX')\overline{\wedge}(D_1D_2X'X)$, so that $(D_1D_2XX')$ is a harmonic range. If two real self-corre-

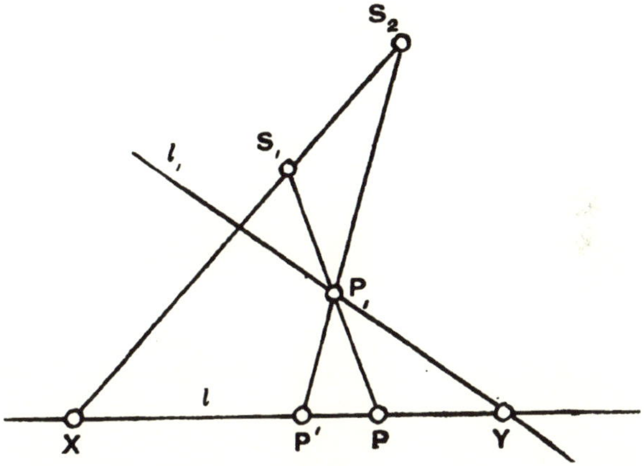


FIG. 59.

sponding points do not exist, we introduce by definition conjugate pairs of "imaginary" points, much in the same way as ideal points were introduced into hyperbolic geometry.

When the double points are real the involution is said to be *hyperbolic*, and when they are imaginary it is said to be *elliptic*. If the double points coincide, the conjugate of any point $P$ coincides with $D$, and the involution is said to be *parabolic*.

If $\{p\}$ and $\{p'\}$ are two homographic pencils with different vertices, the locus of the points of intersection of corresponding lines is a curve with the property that any line cuts it in two points, real, coincident or imaginary. A line $l$

cuts the two pencils in homographic ranges, and the self-corresponding points of these ranges are points on the locus.

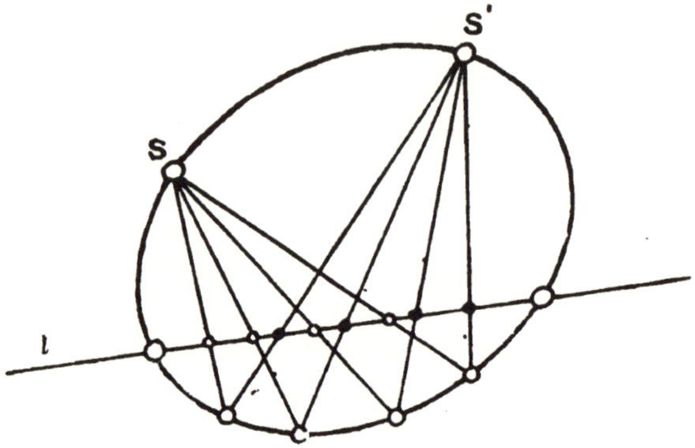


FIG. 60.

This curve is called a *point-conic*; it is the general curve of the second degree, characterised by the property that any line cuts it in two points.

Similarly the envelope of the lines joining pairs of corresponding points on two homographic ranges is a curve of the second class, or *line-conic*, characterised by the property that from any point two tangents can be drawn to it.

It can be proved that a point-conic is also a line-conic, and *vice versa*. The term conic can then be applied to either.

### 6. The absolute.

Let us return now to the absolute polar system in a plane. We shall prove the theorem: *In every polar system in a plane which has the reciprocal property that "if the polar of a point A passes through B, the polar of B passes through A," there is a fixed conic, the locus of points or the envelope of lines which are incident with their polars.*

Consider a line $l$. The polar of a point $P$ on $l$ cuts $l$ in a point $P'$, and the polar of $P'$ passes through $P$. Hence the points of $l$ are connected in pairs and form an involution whose double points are incident with their polars. Every line therefore cuts the locus in two points, and the locus is a point-conic. Similarly the envelope is a line-conic. If $l$ cuts the locus in $P$ and $Q$, the polars of $P$ and $Q$ are lines of the line-conic. Further, the polar of $P$ does not cut the locus in any second point, since the polar of any point upon it passes through $P$; hence the polar of $P$ is a tangent to the point-conic, and the point- and line-conics form one and the same conic.

Similarly, in three dimensions a polar system determines a surface of the second degree or quadric surface.

Applying this theorem to the absolute polar system, we find a conic in a plane or a quadric surface in space which is given absolutely. But as a real point cannot lie on its polar, since it is at the fixed distance of a quadrant from any point of it, this conic or quadric can have no real points.

This imaginary conic, or in space the imaginary quadric surface, is called the *Absolute*.

Let $P$, $P'$ and $Q$, $Q'$ be two pairs of conjugate points on a line $g$, so that $PP' = QQ' =$ a quadrant. Therefore

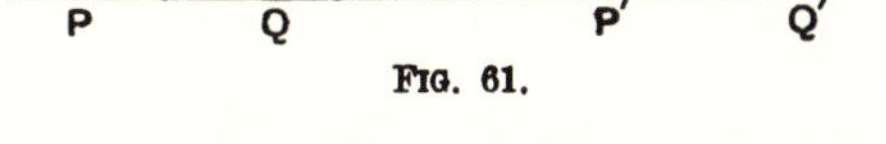


FIG. 61.

$PQ = P'Q'$. Let $g$ cut the absolute in $X$ and $Y$; then $P$, $P'$ and $Q$, $Q'$ are harmonic conjugates with regard to $X$ and $Y$. Let $Q$ coincide with $X$; then $Q'$ will also coincide with $X$, and the equation $PQ = P'Q'$ becomes

$$PX = P'X = PX - PP'.$$

Therefore $1 = 1 - PP'/PX$. Therefore $PX$ must be infinite. Every point on the absolute is therefore at an infinite distance from any real point, and the absolute is, like the real conic in hyperbolic geometry, the locus of points at infinity.

## 7. Principle of duality.

The polar system with regard to the absolute conic establishes the principle of duality. In euclidean geometry the principle of duality holds so long as we are dealing with purely descriptive properties, *i.e.* it holds in projective geometry, which is independent of any hypothesis regarding parallel lines, but it has only a very limited range in metrical geometry, and is often applied more as a principle of analogy than as a scientific principle with a logical foundation.

Thus, four circles can be drawn to touch three given lines, but only one circle can be drawn to pass through three points. A circle is the locus of a point which is always at a fixed distance from a given point, but we cannot consider it also as the envelope of a line which makes a constant angle with a fixed straight line.

In hyperbolic geometry, when we consider equidistant curves as circles, we find it true that four circles are determined by three points; and if we introduce freely points at infinity and ideal points, we can make the principle of duality fit fairly well.

In elliptic geometry, however, the principle of duality has its widest field of validity, and extends to the whole of metrical geometry. The reason for this is found in the nature of the absolute and the measure of distance and angle. In a pencil of lines with vertex $O$ there are always two absolute lines, the tangents from $O$ to the absolute, and in all three geometries these two lines are conjugate imaginaries. They form the double lines of the elliptic involution of pairs of conjugate or rectangular lines through $O$. In a range of points on a line $l$ there are similarly two absolute points, the points of intersection of $l$ with the absolute. They form the double points of the involution of pairs of conjugate points with regard to the absolute. But in hyperbolic, elliptic and euclidean geometry this involution is respectively hyperbolic, elliptic and parabolic. Thus it is only in elliptic geometry that the involution on a line is of the same nature as that in a pencil.

8. As a consequence of this, in elliptic geometry *the distance between two points is proportional to the angle between their absolute polars.*

Consider two lines $OP$, $OQ$. Let $P'$, $Q'$ be the poles of $OP$ and $OQ$. Then $P'Q'$ is the polar of $O$. $PP' = QQ' = q$.

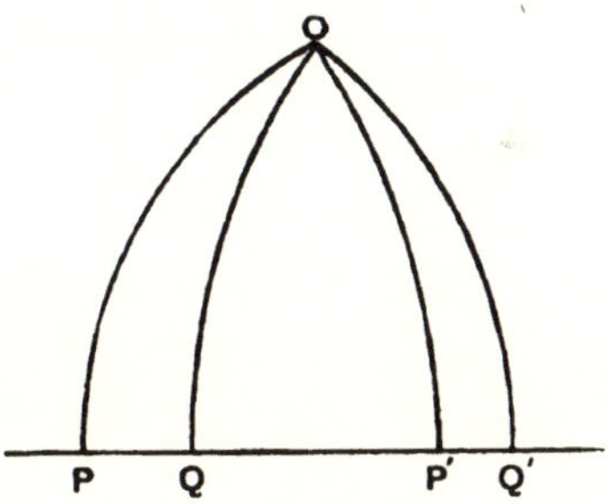


FIG. 62.

Now distances measured along $PQ$ are proportional to the angles at $O$.

Therefore $\dfrac{PQ}{2q} = \dfrac{\angle POQ}{\pi}$ and $P'Q' = PQ$.

Therefore the distance $d$ between the poles of the lines is connected with the angle $a$ between the lines by the relation

$$d = \frac{2q}{\pi} a,$$

and if the unit of distance is such that $q = \frac{\pi}{2}$, then $d = a$.

Here we must observe that two points have two distances, viz. $d$ and $2q - d$; two lines have two angles, $a$ and $\pi - a$. In the above relation we have made the smaller distance correspond to the smaller angle.

Consider, however, a triangle $ABC$, in which we shall suppose each of the sides $< q$, and each of the angles $< \frac{\pi}{2}$.

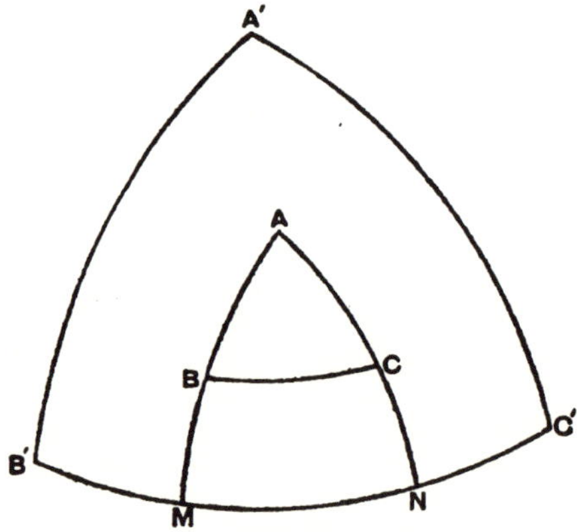


FIG. 63.

The absolute polar figure is another triangle $A'B'C'$, in which $B'C'$ is the polar of $A$, etc. Let $AB$, $AC$ meet $B'C'$ in $M$ and $N$. Then

$$B'N = MC' = q,$$

and $$B'C' = 2q - MN = 2q - \frac{2q}{\pi} A = \frac{2q}{\pi}(\pi - A).$$

To the angle $A$, $<\frac{\pi}{2}$, corresponds therefore the segment $a'$, $>q$.

To a segment $d$ corresponds an angle $\frac{\pi}{2q}(2q-d)$, and to an angle $\alpha$ corresponds a segment $\frac{2q}{\pi}(\pi-\alpha)$. If the segment $d>d'$, the corresponding angle $\alpha<\alpha'$.

In applying the principle of duality, therefore, we must interchange point and line, segment and angle, greater and less.

**9. Area of a triangle.**

Two lines enclose an area proportional to the angle between them, $=2k^2A$, say, where $k$ is a linear constant,

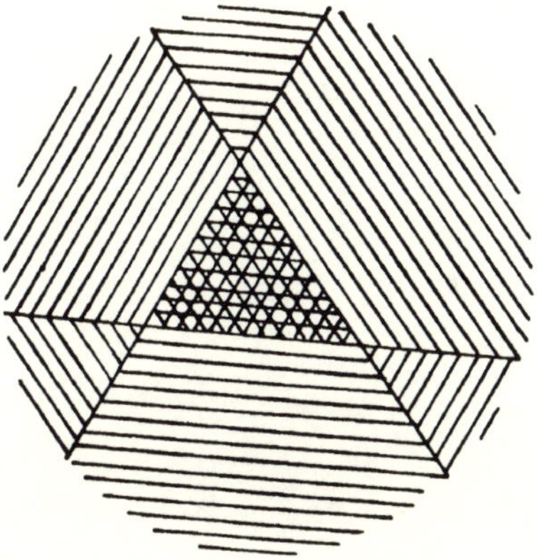

FIG. 64.

and the area of the whole plane is $2k^2\pi$. In Fig. 64 the areas enclosed by the angles of the triangle are shaded, and these areas cover the area of the triangle three times, and the rest of the plane only once.

We have, therefore, $2k^2(A+B+C)=2k^2\pi+2\Delta$,

whence $$\Delta=k^2(A+B+C-\pi).$$

The area of a triangle is therefore proportional to the excess of the sum of its angles over two right angles. If $A_1$, $B_1$, $C_1$ are the exterior angles,

$$\Delta = k^2(2\pi - A_1 - B_1 - C_1).$$

The absolute polar of the triangle $ABC$ is a triangle $A'B'C'$ with sides $a', b', c' = \frac{2q}{\pi}(\pi - A)$, etc. Hence

$$\Delta = k^2 \frac{\pi}{2q}(4q - a' - b' - c'),$$

or *the perimeter of a triangle falls short of* $4q$ *by an amount proportional to the area of the polar triangle.*

These results hold also for the sum of the exterior angles and the perimeter of any simple polygon.

## 10. The circle.

A circle is the locus of points equidistant from a fixed point, the *centre*, and by the principle of duality it is also

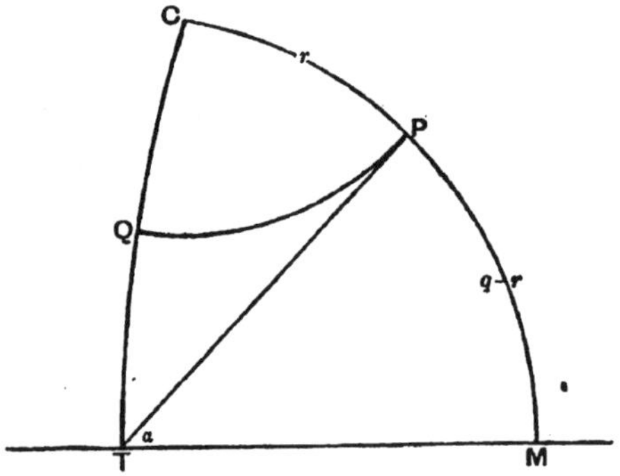


FIG. 65.

the envelope of lines which make a constant angle with a fixed line, the *axis*.

Let $C$ be the centre and $c$ the polar of $C$. Let $P$ be any point on the circle, and draw the tangent $PT$. Then $CM \perp TM$ and also $\perp TP$. Therefore $T$ is the pole of $CP$.

$$PM = q - r = \frac{2q}{\pi} a\text{; therefore } a \text{ is a constant angle} = \frac{\pi}{2} - \frac{\pi}{2q} r.$$

Further, since $PM = q - r$, the circle is an equidistant-curve with $c$ as axis. Just as in hyperbolic geometry, the circle or equidistant-curve lies symmetrically on both sides of the axis, but the two branches are continuously connected.

In elliptic geometry, therefore, equidistant-curves are proper circles. When the radius of a circle is a quadrant the circle becomes a double straight line, the axis taken twice.

**11.** In three dimensions the surface equidistant from a plane is a proper sphere.

A remarkable surface exists which is equidistant from a line. With this property it resembles a cylinder in ordinary space. A section by a plane perpendicular to the axis is a circle. A section by a plane through the axis is an equidistant-curve to the axis, but this is also a circle, and the surface can be generated by revolving a circle about its axis. It thus also resembles an anchor ring (Fig. 66). But sections perpendicular to the axis do not cut it in pairs of circles, but only in single circles, and so it also resembles a hyperboloid of one sheet. Every point is at a distance $d$ from the axis $l$, and is therefore at a distance $q - d$ from the line $l'$, the absolute conjugate of $l$. The surface has therefore two conjugate axes, and can be generated by the revolution of a circle about either of these. It is therefore a double surface of revolution. It is a surface of the second degree, since a straight line cuts it in two points.

From its resemblance to a hyperboloid, the existence of rectilinear generators is suggested. If it does possess rectilinear generators, these lines must be everywhere

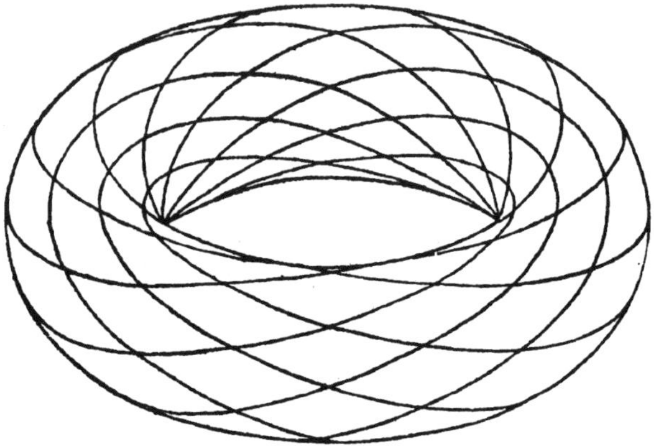

FIG. 66.[1]

equidistant from either axis. We shall therefore investigate the existence of such lines, and return in § 17 to a description of this surface (*Clifford's surface*).

## 12. Common perpendicular to two lines in space.

Consider two lines $a$, $b$ not in the same plane. Let $a'$ and $b'$ be their absolute polars. Any line which cuts both $a$ and $a'$ is perpendicular to both; hence any line which

[1] This picture of Clifford's surface will be best understood after reading Chap. V. In the conformal representation of non-euclidean geometry in euclidean space, planes and spheres are all represented by spheres, straight lines and circles by circles. Clifford's surface is represented by an anchor-ring, one axis being represented by the axis of the ring, the other axis being represented by a line at infinity. The circular sections of the surface by planes through an axis (which are lines of curvature) are represented by the meridians and parallels of the anchor-ring (which are also lines of curvature). The rectilinear generators are represented by the bitangent circular sections of the ring. The two systems of these last-mentioned circles are depicted in the figure. They intersect at a constant angle.

meets the four lines $a$, $b$, $a'$, $b'$ cuts them all at right angles. Now, three of these lines $a$, $b$, $a'$ determine a ruled surface of the second degree, and the fourth line $b'$ cuts this surface in two points $P$, $Q$. The two generators $p$, $q$ of the opposite system through $P$ and $Q$ are common transversals of the four lines $a$, $a'$, $b$, $b'$, and therefore cut all four at right angles. The two lines $a$, $b$ have therefore two common perpendiculars. The two common perpendiculars $p$, $q$ are absolute polars. For, since $a$, $a'$ and $b$, $b'$ cut $p$ and $q$ at right angles, they also cut the polars $p'$ and $q'$, but they have only two common transversals; therefore $p'$ must coincide with $q$, and $q'$ with $p$.

Of the two common perpendiculars one is a minimum and the other a maximum perpendicular from one line on

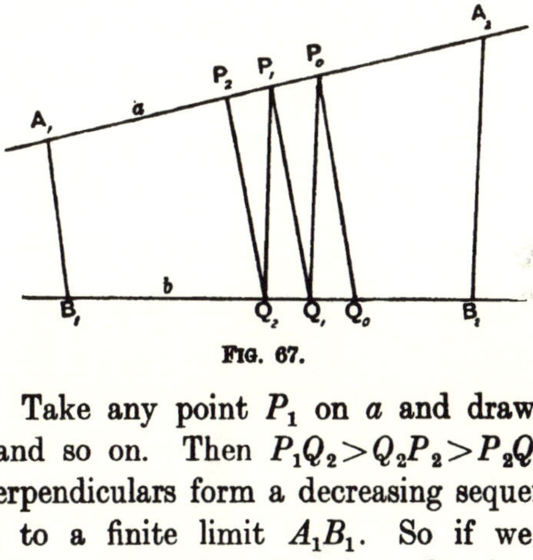


FIG. 67.

the other. Take any point $P_1$ on $a$ and draw $P_1Q_2 \perp b$, $Q_2P_2 \perp a$, and so on. Then $P_1Q_2 > Q_2P_2 > P_2Q_3 > \ldots$, so that the perpendiculars form a decreasing sequence which must tend to a finite limit $A_1B_1$. So if we continue the sequence in the other direction, drawing $P_1Q_1 \perp a$, $Q_1P_0 \perp b$, ..., we have an increasing sequence which tends to the other common perpendicular $A_2B_2$.

**13. Paratactic lines.**

If $A_1B_1 = A_2B_2$, all the intermediate perpendiculars must also be equal; the two lines have then an infinity of common perpendiculars, and therefore the four lines $a$, $b$, $a'$, $b'$ all belong to the same regulus of a ruled surface of the second degree. The two lines are *equidistant, though not coplanar*; they are analogous to parallel lines in ordinary geometry and possesses many of their properties. They were discovered by W. K. Clifford, and have therefore been called *Clifford's parallels*. A more distinctive name, suggested by Study, is *paratactic* lines.

Through any point $O$ two lines can be drawn paratactic to a given straight line, one right-handed and the other left-handed. Each is obtained from the original line by screwing it along the perpendicular $NO$ either right-handedly or left-handedly. The angle through which it has to be turned is proportional to the distance through which it has to be moved.

In the plane $ONM$ draw $OM \perp ON$, cutting the given line in $M$, the pole of $ON$ in this plane. Draw $MP$ the

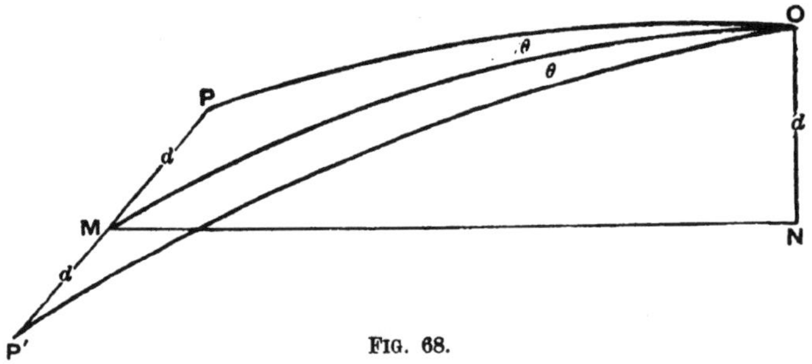


FIG. 68.

polar of $ON$, which is therefore perpendicular to $NM$, and along it cut off $MP = MP' = d$. Then $OP$ and $OP'$ are the two lines through $O$ paratactic to $NM$. Also $d = \frac{2q}{\pi}\theta$.

**14.** The above construction for a common perpendicular to two skew lines can only be carried out in elliptic geometry, for in hyperbolic geometry the polar of a real point is ideal, and in euclidean geometry it is at infinity. Consider a system of pairs of planes at right angles to each other

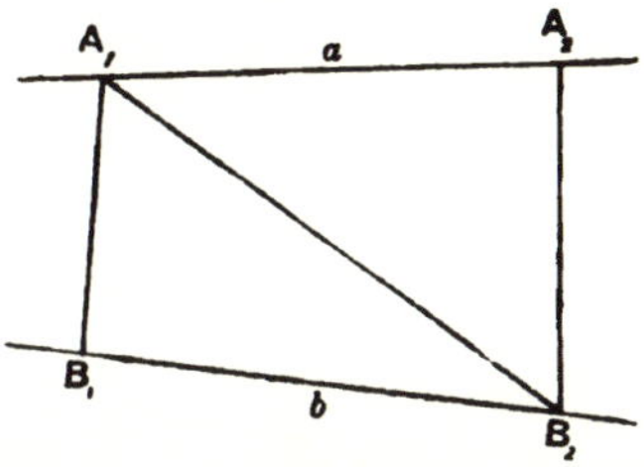


FIG. 69.

drawn through the line $a$. This forms an elliptic involution, the double elements of which are the imaginary planes through $a$ which touch the absolute. These planes are cut by the line $b$ in a range of points forming an elliptic involution. Although the double points of this involution are imaginary, the centres $B_1$, $B_2$ of the segments determined by the double points are always real. These form a pair of elements of the involution a quadrant apart. In elliptic geometry there are two real centres, in euclidean geometry one is at infinity, and in hyperbolic geometry one is ideal. The perpendiculars to $b$ at $B_1$, $B_2$ are the two common perpendiculars.

Since $B_1$ and $B_2$ are conjugate points and $A_1B_1 \perp B_1B_2$, $A_1B_1$ is the polar of $B_2$ in the plane $A_1b$; therefore $A_1B_2 \perp A_1B_1$. But the plane $aB_1 \perp$ the plane $aB_2$; therefore $A_1B_1$ is $\perp$ the plane $aB_2$; therefore $A_1B_1 \perp a$. Similarly $A_2B_2 \perp a$.

The points $A_1$, $A_2$ are the centres of a similar elliptic involution on $a$.

**15.** *Two paratactic lines cut the same two generators, of the same system, of the absolute.*

Let three common transversals $l_1$, $l_2$, $l_3$ cut the four lines $a$, $a'$, $b$, $b'$ in $A_i$, $A_i'$, $B_i$, $B_i'$, and the absolute in $X_i$, $Y_i$ ($i=1, 2, 3$). Then, since $A_1$, $A_1'$ and $B_1$, $B_1'$ are

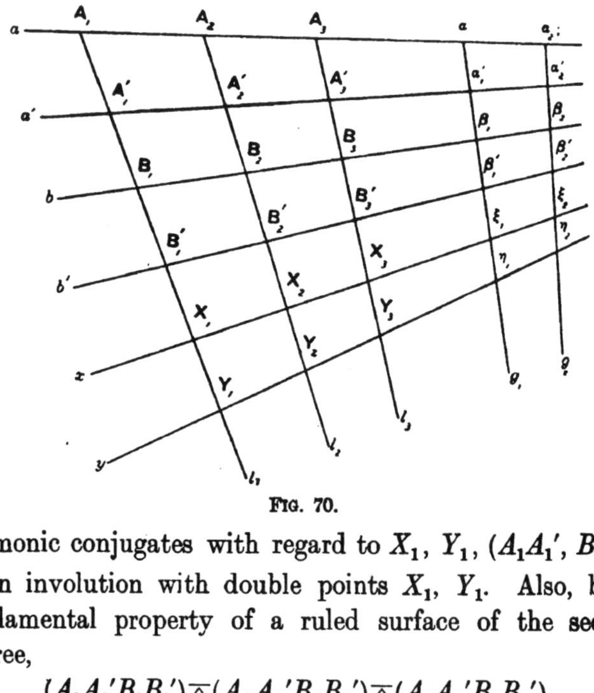

FIG. 70.

harmonic conjugates with regard to $X_1$, $Y_1$, $(A_1A_1', B_1B_1')$ is an involution with double points $X_1$, $Y_1$. Also, by a fundamental property of a ruled surface of the second degree,

$$(A_1A_1'B_1B_1') \barwedge (A_2A_2'B_2B_2') \barwedge (A_3A_3'B_3B_3').$$

Therefore

$$(A_1A_1'B_1B_1'X_1Y_1) \barwedge (A_2A_2'B_2B_2'X_2Y_2)$$
$$\barwedge (A_3A_3'B_3B_3'X_3Y_3).$$

Therefore $X_1X_2X_3$ and $Y_1Y_2Y_3$ are two generators $x$, $y$ of the same regulus as $A_1A_2A_3$, etc.; they are also generators of the same regulus of the absolute, since they cut it in more than two points. Hence *all the common perpendiculars to two paratactic lines cut the same two generators of the absolute.*

Let $a$ cut the absolute in $a_1, a_2$. Through each of these points passes one generator ($g_1$ and $g_2$) of the absolute of the opposite system to $x$, $y$, and therefore cutting $x$, $y$ in $\xi_1\xi_2, \eta_1\eta_2$. $g_1$ and $g_2$ must also belong to the same regulus as $l_1$, $l_2$, $l_3$, since they cut the surface in more than two points. Therefore they cut $a'$, $b$ and $b'$ also. Hence $a$ and $b$ cut the same two generators of the absolute. Q.E.D.

Conversely, if $a$ and $b$ cut two generators $g_1, g_2$ of the absolute in $a_1\beta_1, a_2\beta_2$, let $h_1$, $h_2$ be the two generators of the other system through $a_1, a_2$, then the polar of $a$ is the intersection of the planes $(g_1h_1)$, $(g_2h_2)$, and therefore cuts both $g_1$ and $g_2$. Hence $g_1$ and $g_2$ are common transversals of $a$, $a'$, $b$, $b'$. But, by § 12, if $a$, $a'$, $b$, $b'$ do not all belong to the same regulus, they have only two common transversals, which are absolute polars. Now $g_1$ and $g_2$ are not absolute polars (each being its own polar), hence $a$, $a'$, $b$, $b'$ belong to the same regulus, and have an infinity of common transversals. Therefore $a$, $b$ are paratactic.

The two sets of generators of the absolute may be called *right-handed* and *left-handed*. Two lines which cut the same two left (right) generators of the absolute are called left (right) paratactic lines. We see, therefore, that *all the common transversals of two right paratactic lines are left paratactic lines.* Further, if $a$ and $b$ are both right (left) paratactic to $c$, then $a$ is right (left) paratactic to $b$; for $a$, $b$, $c$ all cut the same two generators of the absolute.

**16.** Paratactic lines have many of the properties of ordinary euclidean parallels. In particular they have the characteristic property of being equidistant. They are not, however, coplanar. We shall use the symbol $\sqcap$ for right parataxy, and $\sqcup$ for left parataxy.

If $AB \sqcap CD$ and if $AC$ and $BD$ are both $\perp CD$, they are also $\perp AB$; $AC = BD$ and $\sqcup BD$, and $AB = CD$. Also $AD$ cuts both pairs of lines at equal angles. The figure $ABDC$ is a skew rectangle; its opposite sides are equal and paratactic.

If $AB \sqcap CD$ and $= CD$, then joining $AC$, $BD$ and $AD$, $\angle ADC = \angle DAB$, and we find two congruent triangles $ACD$ and $DBA$; therefore

$$AC = BD \quad \text{and} \quad \angle ACD = \angle DBA.$$

Conversely, if $AB = CD$ and $\angle ABD = \angle DCA$, or if $CAB + ACD = 2$ right angles, then $AB$ is paratactic to $CD$. Hence, if $AB =$ and $\sqcap CD$, it follows that $AC =$ and $\sqcup BD$. $ABDC$ is analogous to a parallelogram.

Real parataxy can only exist in elliptic space. For if $ABDC$ is a skew rectangle, the lines $AB$, $AC$, $AD$ are not co-planar.

Therefore $\angle CAD + BAD > \angle CAB$, *i.e.* $>$ a right angle.

But $$\angle BAD = \angle ADC;$$

therefore $$\angle CAD + ADC + ACD > 2 \text{ right angles};$$

therefore the geometry is elliptic.

### 17. Clifford's surface.

If $a \sqcap c$ and $b \sqcap c$, so that $a \sqcap b$, the common transversals of $a$, $b$, $c$ are all $\sqcup$, and form one regulus of a ruled surface of the second degree; the lines of the other regulus are all $\sqcap$ $a$, $b$ and $c$. If $a'$ is a generator of the opposite system to $a$, $b$, $c$, then any line which cuts $a$ and is $\sqcup$ $a'$ cuts $b$ and $c$. The surface is therefore generated by a line which cuts a fixed line and is paratactic to another fixed line. By § 16 it cuts the fixed line at a constant angle, $2\theta$.

Let *OP* be the fixed line, which is cut by the variable line *OP'* (Fig. 68). Draw $ON \perp$ the plane *POP'* and $= d = 2\theta q/\pi$. Draw *OM* bisecting the angle *POP'* ($=2\theta$), and draw $NM \perp ON$ in the plane *MON*. *NM* is then paratactic to both *OP* and *OP'*, and *ON* is supposed to be drawn in the direction such that $NM \sqcap OP$ and $\sqcup OP'$. Any other line which cuts *OP* and is $\sqcup OP'$, *i.e.* any generator of the left-handed system, is also $\sqcup NM$, and is at the same distance *d* from *NM*. Hence *NM* is an axis of revolution of the surface, and similarly the polar of *NM* is also an axis of revolution.

This surface is, therefore, just the surface of revolution of a circle about its axis which we considered in § 11. In fact, through any point *P* of this surface there pass two lines paratactic to the axis, and since these lines are equidistant from the axis, they lie entirely in the surface. This surface, which is called *Clifford's Surface*, is therefore a ruled surface. All the generators of one set are $\sqcap$ to the axis $l$, and all the generators of the other set are $\sqcup l$. Two generators of opposite systems cut at a constant angle $\frac{\pi}{q} d$. From the figure in § 15, it appears that Clifford's surface cuts the absolute in two generators of each system.

Suppose the surface is cut along two generators. The whole surface is covered with a network of lines intersecting at a fixed angle $2\theta$, and can be conformly represented upon a euclidean rhombus with this angle. The geometry on this surface is therefore exactly the same as that upon a finite portion of the euclidean plane bounded by a rhombus whose opposite sides are to be regarded as coincident. As an immediate consequence, the area of the surface is found to be $4q^2 . \sin 2\theta$, since the side of the rhombus $= 2q$. We have therefore the remarkable result

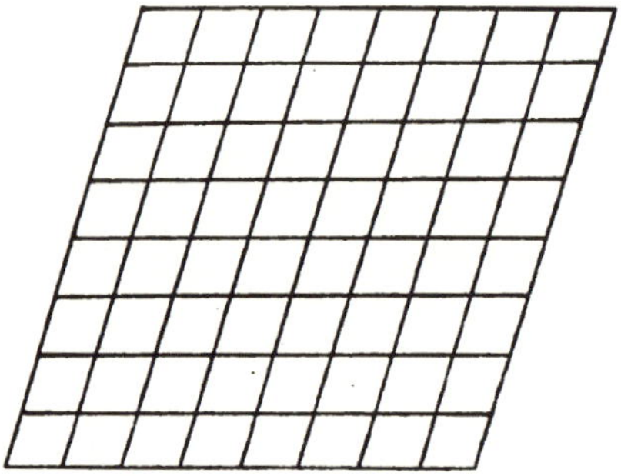

FIG. 71.

that both in hyperbolic and in elliptic space there exist surfaces (viz. horospheres and Clifford's surfaces respectively) upon which euclidean geometry holds.

## 18. Trigonometrical formulae. Circumference of a circle.

In investigating the trigonometrical formulae we shall use a method which might equally well have been employed in hyperbolic geometry. The starting point is the assumption that euclidean geometry holds in the infinitesimal domain.[1]

[1] The truth of this assumption is indicated by the fact that when the sides of a triangle tend to zero, the sum of the angles tends to the value $\pi$. The steps of the proof are as follows. Let $ABC$ be a triangle with right angle at $C$. We have to prove (1) that the ratio $AC : AB$ tends to a limit. This limit is a function of the angle $A$, say $f(A)$. We have to prove (2) that $f(A)$ is continuous, and (3) that its value is $\cos A$. The last step is best obtained by the formation of a functional equation $f(\theta+\phi)+f(\theta-\phi)=2f(\theta) . f(\phi)$. See Coolidge, Chap. IV.

Let $AOB$ be a small angle $a$, $OA = OB = r$, $AA' = BB' = dr$, $AB = a$, $A'B' = a + da$. The angle $OAB$ is nearly $= \frac{\pi}{2}$. Let $OAB = OBA = \frac{\pi}{2} - \theta$, $OA'B' = OB'A' = \frac{\pi}{2} - (\theta + d\theta)$.

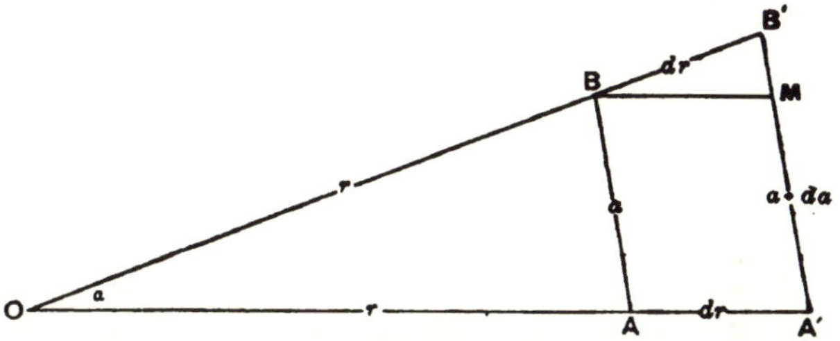


FIG. 72.

Draw $BM$, making the angle $ABM = ABO = \frac{\pi}{2} - \theta$.

Then, neglecting higher infinitesimals, we have $A'M = AB$; therefore $MB' = da$, and $BM = dr$.

Therefore $\quad da = 2dr \sin \frac{1}{2} MBB' = 2dr \sin \theta$,

or
$$\frac{da}{dr} = 2\theta.$$

Again, the area $ABB'A' = a\,dr$, and the sum of its exterior angles
$$= 2\left(\frac{\pi}{2} - \theta + \frac{\pi}{2} + \theta + d\theta\right) = 2(\pi + d\theta);$$

therefore
$$a\,dr = -2k^2 d\theta, \quad (\S\ 9)$$

or
$$a = -2k^2 \frac{d\theta}{dr} = -k^2 \frac{d^2a}{dr^2},$$

*i.e.*
$$\frac{d^2a}{dr^2} + \frac{a}{k^2} = 0.$$

The solution of this is
$$a = C \sin\left(\frac{r}{k} + \phi\right).$$

Differentiating, $\displaystyle \frac{da}{dr} = 2\theta = \frac{C}{k} \cos\left(\frac{r}{k} + \phi\right).$

When $r=0$, $a=0$ and $2\left(\frac{\pi}{2}-\theta\right)=\pi-\alpha$, therefore $2\theta=\alpha$, so that $0=C\sin\phi$, $\alpha=\frac{C}{k}\cos\phi$; whence $\phi=0$ and $C=k\alpha$.

Hence we have, finally,

$$a=k\alpha\sin\frac{r}{k}.$$

Since $a$ and $\alpha$ are small, we can take $a$ as the arc of a circle of radius $r$. The whole circumference of a circle is therefore $2\pi k\sin\frac{r}{k}$.

When $r=\frac{1}{2}\pi k$, the circumference is $2\pi k$, which is twice the length of a complete line, and therefore $q=\frac{1}{2}\pi k$.

**19. Trigonometrical formulae for a right-angled triangle.**

Keep one part, say $b$, fixed. Let $BAB'=dA$, $BB'=da$,

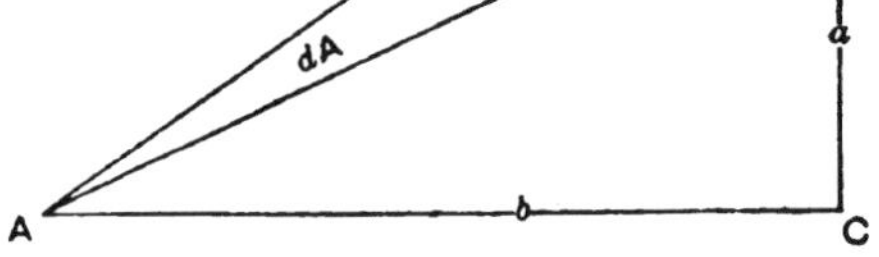


FIG. 73.

$AB'C=B+dB$, $NB'=dc$.

Then $$dc=da\cos B, \qquad\text{.........(1)}$$

$$k\sin\frac{c}{k}\,dA=NB=da\sin B. \qquad\text{.........(2)}$$

The area of $BAB'$ is obtained in two ways, (1) by integrating

$$= \int_0^c k \sin \frac{c}{k}\, dA\, dc$$

$$= k^2 dA \left(1 - \cos \frac{c}{k}\right);$$

(2) in terms of the angular excess

$$= k^2(dA + \pi - B + B + dB - \pi)$$
$$= k^2(dA + dB).$$

Equating these, $$dB = -\cos \frac{c}{k}\, dA. \qquad \text{.........................(3)}$$

Eliminate $da$ and $dA$ between these three equations, and we get

$$k\, dB \tan \frac{c}{k} = -k \sin \frac{c}{k}\, dA = -da \sin B = -dc \tan B,$$

giving a differential equation in $B$ and $c$. The integral of this is

$$\sin \frac{c}{k} \sin B = f(b),$$

since $b$ is the only constant part.

Putting $B = \frac{\pi}{2}$, $c = b$, and we find $f(b) = \sin \frac{b}{k}$.

Hence we have $\sin \frac{b}{k} = \sin \frac{c}{k} \sin B.$

**20. Associated triangles.**

In order to obtain the other relations between the sides and angles, we shall establish a sequence of associated triangles which form the basis for Napier's rules in spherical trigonometry. This sequence has already been referred to, and a similar sequence was found in hyperbolic geometry.

We shall first introduce the following notation. Let $\alpha = a/k$, $\alpha' = \frac{\pi}{2} - \alpha$; then the angles $\alpha$, $\beta$, $\gamma$ correspond to the sides $a$, $b$, $c$ of the triangle.

Draw the absolute polars of the vertices $A$ and $B$. These form, with the sides produced of the given triangle, a star

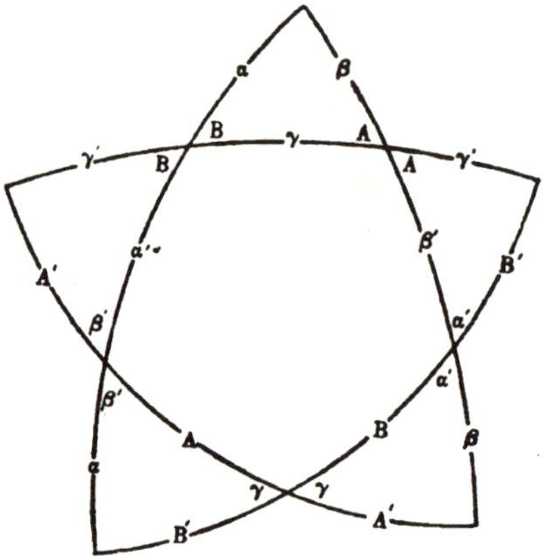

FIG. 74.

pentagon. Mark on each side the angle which corresponds to it, and we get the figure (Fig. 74). Each of the five outer angles is a right angle. Each vertex of the simple pentagon is the pole of the opposite side. We obtain then five associated right-angled triangles. If we write down the five quantities $A$, $\alpha'$, $\gamma$, $\beta'$, $B$ which correspond to the parts of the first triangle $A$, $\alpha$, $\gamma$, $\beta$, $B$, the corresponding quantities in the same order for the second triangle are $\alpha'$, $\gamma$, $\beta'$, $B$, $A$, but these are the same as the five quantities corresponding to the first triangle permuted cyclically; and they are represented in proper order by the sides of the simple pentagon.

**21. Napier's rules.**

Now we have proved for the first triangle that

$$\sin\frac{b}{k}=\sin\frac{c}{k}\sin B.$$

Writing this in terms of $A$, $\alpha'$, $\gamma$, $\beta'$, $B$, we have

$$\cos\beta'=\sin\gamma\sin B,$$

and since this equation can be applied to each of the five triangles, and therefore transformed by cyclic permutation, we can state a general rule as follows:

Write the five angles $A$, $\alpha'$, $\gamma$, $\beta'$, $B$ in order on the sides of a simple pentagon. Then, calling any one part the middle part and the other two pairs the adjacent parts and the opposite parts, we have

cos (middle part) = product of sines of adjacent parts. (A)

Taking in succession $\gamma$, $A$, $B$ as middle parts, we get

$$\cos\gamma=\sin\alpha'\sin\beta',$$
$$\cos A=\sin\alpha'\sin B,$$
$$\cos B=\sin\beta'\sin A.$$

Hence $$\cos\gamma=\cot A\cot B,$$

*i.e.* cos (middle part) = product of cotangents of opposite parts. ....................................(B)

There is a relation of one of these forms between any three parts of the triangle. For convenience we write down the ten relations in terms of $a$, $b$, $c$, $A$, $B$.

$$\cos\frac{c}{k}=\cos\frac{a}{k}\cos\frac{b}{k}=\cot A\cot B,$$

$$\sin\frac{a}{k}=\sin\frac{c}{k}\sin A=\tan\frac{b}{k}\cot B,$$

$$\cos A=\cos\frac{a}{k}\sin B=\cot\frac{c}{k}\tan\frac{b}{k},$$

and two other pairs formed by interchanging $a$, $b$ and $A$, $B$.

These are exactly the same as the relations which exist between the parts of a spherical triangle. *The trigonometry of the elliptic plane is therefore exactly the same as ordinary spherical trigonometry.*

If we write the parts $\alpha$, $B'$, $\gamma'$, $A'$, $\beta$ in the order in which they occur in the triangle, we get the more familiar rules of Napier:

sine (middle part) = product of cosines of opposite parts
= product of tangents of adjacent parts.

**22.** In elliptic space *the formulae for spherical trigonometry are the same as in euclidean space*, when we take as the measure of a side of a spherical triangle the angle which it subtends at the centre, and as the measure of an angle the dihedral angle between the planes passing through the sides and the centre.

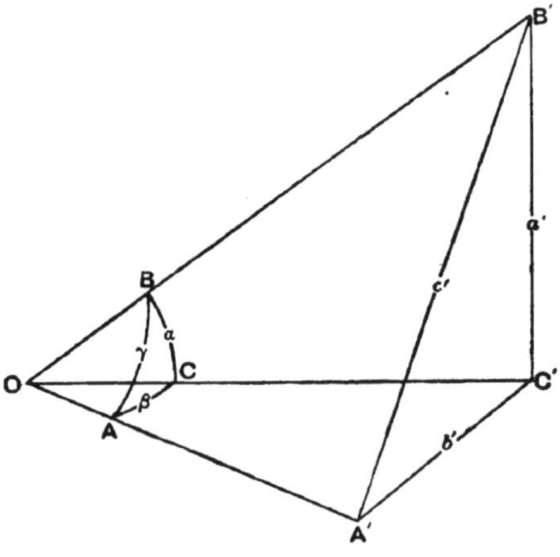


FIG. 75.

Let $O$ be the centre of the sphere, and let $OA$, $OB$, $OC$ cut the polar plane of $O$ in $A'$, $B'$, $C'$. Then we get a rectilinear triangle $A'B'C$ with sides $a'$, $b'$, $c'$. The angles which the radii $OA'$, etc., make with the sides are right angles; hence $A'$ = the dihedral angle between the planes $OAB$ and $OAC$, *i.e.* $A' = A$. Also $a' = ka$. Hence the relations between $a$, $\beta$, $\gamma$, $A$, $B$, $C$ are the same as those between

$\frac{a'}{k}, \frac{b'}{k}, \frac{c'}{k}$, $A'$, $B'$, $C'$, which are the same as those of ordinary spherical trigonometry.

The measurement of angle, plane or dihedral, is the same in all three kinds of space, and spherical trigonometry involves only angular measurement. This explains why spherical trigonometry is the same in all three geometries.

## 23. The trirectangular quadrilateral.

As in hyperbolic geometry, there is a correspondence between a right-angled triangle and a trirectangular quadrilateral. In fact

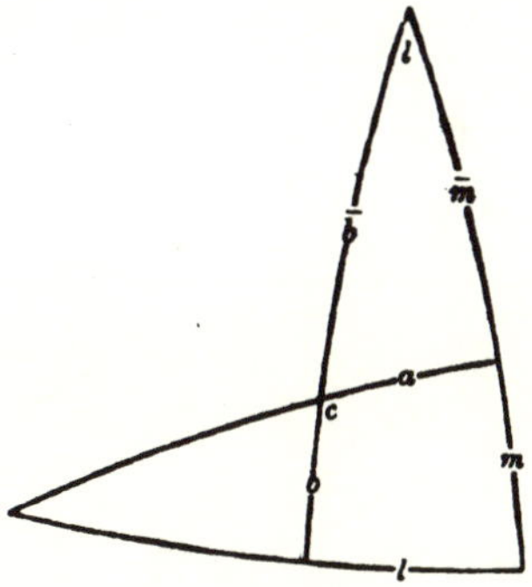


FIG. 76.

we see that, by producing two opposite sides of the quadrilateral to meet, we get, corresponding to the trirectangular quadrilateral *Camlb*, a right-angled triangle with hypotenuse $\frac{1}{2}\pi k - b = \bar{b}$, sides $a$ and $\frac{1}{2}\pi k - m = \bar{m}$, and the opposite angles $l$ and $\pi - C$.

If we write the parts

$$\frac{\pi}{2}+C,\quad \frac{a}{k},\quad \frac{\pi}{2}-\frac{m}{k},\quad \frac{\pi}{2}-\frac{l}{k},\quad \frac{b}{k}$$

in cyclic order, then we have the rules:

sine (middle part) = product of cosines of opposite parts
= product of tangents of adjacent parts.

## EXAMPLES III.

1. Prove that the bisectors of the vertical angle of a triangle divide the base into segments whose sines are in the ratio of the sines of the sides.

2. Prove that the arc of an equidistant-curve of distance $a$, corresponding to a segment $x$ on its axis, is $x \cos a/k$.

3. Prove that the area of a circle of radius $r$ is $4\pi k^2 \sin^2 \dfrac{r}{2k}$.

4. Prove that the area included between an arc of an equidistant-curve of distance $a$, its axis, and two ordinates at distance $x$, is $kx \sin \dfrac{a}{k}$.

5. Prove that the area of the whole plane is $2\pi k^2$, and the volume of the whole of space is $\pi^2 k^3$.

6. Prove that the volume of a sphere of radius $r$ is

$$\pi k^3\left(\frac{2r}{k} - \sin\frac{2r}{k}\right).$$

(In the following examples $k$ is unity.)

7. If $R$ is the radius of the circumsphere of a regular tetrahedron whose side is $a$, show that

$$\sin \tfrac{1}{2}a = \sqrt{\tfrac{2}{3}} \sin R.$$

8. If $2d$ is the distance between opposite edges of a cube of edge $2a$, $2h$ the distance between opposite faces, and $R$ the radius of the circumsphere, prove that

$$\sin^2 d = 2\tan^2 a,\quad \sin^2 h = \sin^2 a/\cos 2a,\quad \sin^2 R = 3\sin^2 a.$$

9. A semiregular network is formed of triangles and quadrilaterals, two of each at each node. Prove that this can only exist in elliptic space, and that the length of the side is $\tfrac{1}{3}\pi$.

**10.** In elliptic geometry show that there can exist six equal circles, each touching each of the others, and of radius given by $2\cos r \sin\frac{\pi}{5} = 1$; three equal circles each having double contact with the other two, and of radius $\frac{\pi}{4}$; and (with overlapping) four circles each touching the other three, and of radius $\cos^{-1}\frac{1}{\sqrt{3}}$.

**11.** Prove that five spheres, each of radius $\frac{\pi}{3}$, can be placed each touching the other four; eight spheres, each of radius $\frac{\pi}{4}$, and each having double contact with four others; and four spheres of radius $\frac{\pi}{4}$, each having double contact with the other three.

**12. For a regular polyhedron:**

$a$ = length of edge.
$\alpha$ = angle subtended by edge at centre.
$\theta$ = angle of each polygon.
$\delta$ = dihedral angle between faces.
$n$ = number of sides of each face.
$p$ = number of edges at each vertex.
$R$ = radius of circumscribed sphere.
$r$ = radius of inscribed sphere.
$\rho$ = radius of sphere touching the edges.
$R_a$ = radius of circumcircle of each face.
$r_a$ = radius of incircle of each face.

Prove the relations:

$$\cos\frac{\pi}{n} = \sin\frac{\pi}{p}\cos\frac{\alpha}{2}, \quad \sin\frac{a}{2} = \sin R \sin\frac{\alpha}{2}, \quad \sin\rho = \tan\frac{a}{2}\cot\frac{\alpha}{2},$$

$$\cos\frac{\pi}{n} = \sin\frac{\theta}{2}\cos\frac{a}{2}, \quad \sin\frac{a}{2} = \sin R_a \sin\frac{\pi}{n}, \quad \sin r_a = \tan\frac{a}{2}\cot\frac{\pi}{n},$$

$$\cos R = \cos r \cos R_a, \quad \cos\rho = \cos r \cos r_a, \quad \sin r_a = \tan r \cot\frac{\delta}{2},$$

$$\sin r = \sin\rho \sin\frac{\delta}{2}. \quad \sin^2\frac{\delta}{2} = \cos^2\frac{a}{2}\cos^2\frac{\pi}{p}\Big/\left(\cos^2\frac{a}{2} - \cos^2\frac{\pi}{n}\right).$$

**13.** For a regular tetrahedron prove that $\cos\delta = \cos a/(1+2\cos a)$.

" " hexahedron " $\cos\delta = (\cos a - 1)/2\cos a$

" " octahedron " $\cos\delta = -1/(1+2\cos a)$.

" " dodecahedron prove that

$$\cos\delta = \{2\cos a - (1+\sqrt{5})\}/\{4\cos a + (1-\sqrt{5})\}.$$

For a regular icosihedron prove that

$$\cos\delta = \{(1-\sqrt{5})\cos a - (1+\sqrt{5})\}/2(1+2\cos a).$$

**14.** Prove that elliptic space can be filled twice over by 5 regular tetrahedra of side $\cos^{-1}(-\frac{1}{4})$, with 3 at each edge and 4 at each vertex.

**15.** Prove that elliptic space can be filled in the following ways:

(1) 4 cubes, of edge $\frac{\pi}{3}$, 3 at each edge and 4 at each vertex.

(2) 8 tetrahedra, of edge $\frac{\pi}{2}$, 4 at each edge and 8 at each vertex.

(3) 12 octahedra, of edge $\frac{\pi}{3}$, 3 at each edge and 6 at each vertex.

(4) 60 dodecahedra, of edge $\cos^{-1}\frac{1+3\sqrt{5}}{8}$, 3 at each edge and 4 at each vertex.

(5) 300 tetrahedra, of edge $\cos\frac{\pi}{5}$, 5 at each edge and 20 at each vertex.

# CHAPTER IV.

## ANALYTICAL GEOMETRY.

### 1. Coordinates.

We shall assume elliptic geometry as the standard case, and construct a system of coordinates. The formulae can be adapted immediately to hyperbolic geometry by changing the sign of $k^2$.

Take two rectangular axes $Ox$, $Oy$. Let $P$ be any point, and draw the perpendiculars $PM = u$ and $PN = v$. Let $OP = r$, $xOP = \theta$.

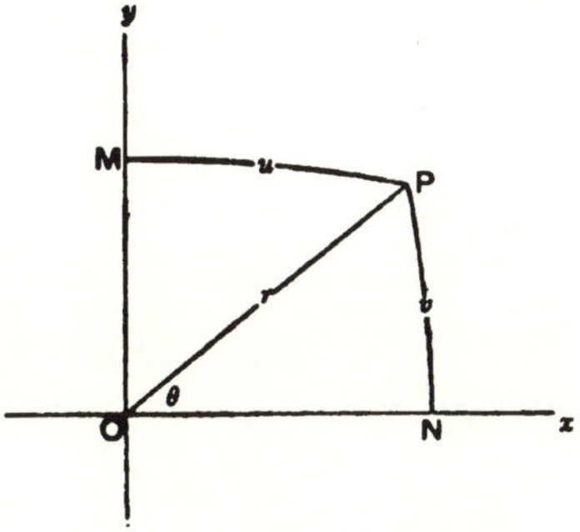


FIG. 77.

$r$, $\theta$ are the *polar coordinates* of the point. $u$, $v$ might be taken as rectangular coordinates, but we shall find it more convenient to take certain functions of these.

We have $\quad \sin\frac{u}{k} = \sin\frac{r}{k}\cos\theta, \quad \sin\frac{v}{k} = \sin\frac{r}{k}\sin\theta.$

For any point on $OP$, therefore, $\sin\frac{v}{k}=\sin\frac{u}{k}\tan\theta$. This is the equation of $OP$ in terms of the coordinates $u$ and $v$.

Consider any line. Draw the perpendicular $ON=p$, and let $xON=\alpha$. $p$ and $\alpha$ are always real, and completely

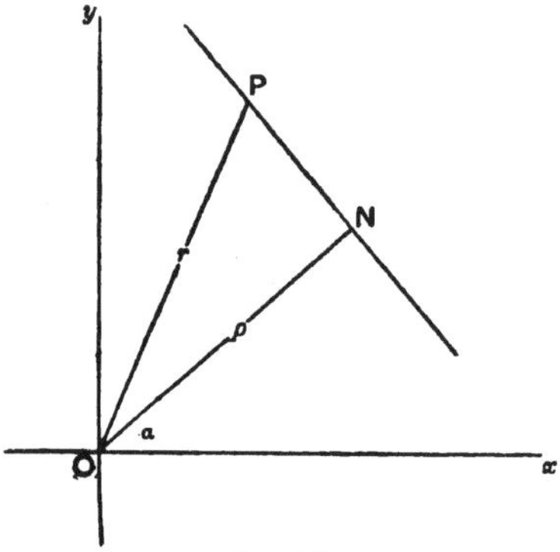


FIG. 78.

determine the line. If $P$ is any point on the line with coordinates $u$, $v$,

$$\tan\frac{p}{k}\cot\frac{r}{k}=\cos(\theta-\alpha).$$

Therefore $$\tan\frac{p}{k}\cos\frac{r}{k}=\sin\frac{u}{k}\cos\alpha+\sin\frac{v}{k}\sin\alpha.$$

This equation is linear and homogeneous in

$$\sin\frac{u}{k},\quad \sin\frac{v}{k},\quad \cos\frac{r}{k}.$$

We shall effect a great simplification, therefore, if we take as coordinates certain multiples of these functions. The equation of a straight line being now of the first degree, the degree of any homogeneous equation in these coordinates gives the number of points in which a straight line meets the curve, *i.e.* the degree of the equation is the same as the degree of the curve.

In order that the coordinates of a real point may be real numbers, both in elliptic and in hyperbolic geometry, we shall define the coordinates as follows:

$$x = k \sin \frac{u}{k} = k \sin \frac{r}{k} \cos \theta,$$

$$y = k \sin \frac{v}{k} = k \sin \frac{r}{k} \sin \theta,$$

$$z = \cos \frac{r}{k}.$$

These are called *Weierstrass' point-coordinates.*

The three homogeneous coordinates are connected by a fixed relationship. We have

$$x^2 + y^2 = k^2 \sin^2 \frac{r}{k} = k^2(1 - z^2),$$

*i.e.*

$$x^2 + y^2 + k^2 z^2 = k^2.$$

As any equation in $x$, $y$, $z$ may be made homogeneous by using this identical relation, we need only, in general, use the ratios of the coordinates.

## 2. The absolute.

In hyperbolic geometry, putting $ik$ instead of $k$, we find the coordinates

$$x = k \sinh \frac{u}{k},$$

$$y = k \sinh \frac{v}{k},$$

$$z = \cosh \frac{r}{k},$$

and $x$, $y$, $z$ are connected by the relationship

$$x^2 + y^2 - k^2 z^2 = -k^2.$$

If $r$ is infinite, $x$, $y$, $z$ are all infinite, but they have definite limiting ratios. Let $\alpha$, $\beta$, $\gamma$ be the actual values, $x, y, z$ the ratios, so that $\alpha=\lambda x$, $\beta=\lambda y$, $\gamma=\lambda z$, and $\lambda\to\infty$.

Then $$\alpha^2+\beta^2-k^2\gamma^2=-k^2;$$

therefore $$x^2+y^2-k^2z^2=-\frac{k^2}{\lambda^2}=0.$$

Hence the *ratios* of the coordinates of a point at infinity satisfy the equation

$$x^2+y^2-k^2z^2=0.$$

This is the equation of the absolute, which is therefore a curve of the second degree or a conic. In hyperbolic geometry it is a real curve; in elliptic geometry the equation is $x^2+y^2+k^2z^2=0$, which represents an imaginary conic.

**3. Normal form of the equation of a straight line. Line-coordinates.**

We found the equation of a straight line in terms of the perpendicular $p$ and the angle $\alpha$, which this perpendicular makes with the $x$-axis, in the form

$$x\cos\alpha+y\sin\alpha=kz\tan\frac{p}{k},$$

which may be written

$$\xi x+\eta y+\zeta z=0.$$

The ratios $\xi:\eta:\zeta$ determine the line, and can be taken as its line-coordinates. It is convenient to take certain multiples of these as actual homogeneous coordinates, viz.

$$\xi=\cos\alpha\cos\frac{p}{k},$$

$$\eta=\sin\alpha\cos\frac{p}{k},$$

$$\zeta=-k\sin\frac{p}{k},$$

which are connected by the identical relation

$$k^2\xi^2+k^2\eta^2+\zeta^2=k^2.$$

These are called *Weierstrass' line-coordinates.*

In hyperbolic geometry

$$\xi=\cos a\cosh\frac{p}{k},\quad \eta=\sin a\cosh\frac{p}{k},\quad \zeta=-k\sinh\frac{p}{k},$$

and the identical relation is

$$k^2\xi^2+k^2\eta^2-\zeta^2=k^2.$$

If $p\rightarrow\infty$, $\xi$, $\eta$, $\zeta$ all $\rightarrow\infty$. Let the actual values be $a$, $\beta$, $\gamma$, and let $a=\lambda\xi$, $\beta=\lambda\eta$, $\gamma=\lambda\zeta$; then

$$k^2(\xi^2+\eta^2)-\zeta^2=k^2/\lambda^2=0.$$

Hence the coordinates of a line at infinity satisfy the equation

$$k^2\xi^2+k^2\eta^2-\zeta^2=0.$$

A homogeneous equation in line-coordinates $\xi$, $\eta$, $\zeta$ represents an envelope of lines. This equation represents an envelope of class 2, *i.e.* a conic. This is the same conic as we had before and represents the absolute, since it expresses the condition that the line $(\xi, \eta, \zeta)$ should be a tangent to $x^2+y^2-k^2z^2=0$.

**4. Distance between two points.**

Let $P(x, y, z)$ and $P'(x', y', z')$ be the two points, $PP'=d$. Then, if the polar coordinates are $(r, \theta)$ and $(r', \theta')$,

$$\cos\frac{d}{k}=\cos\frac{r}{k}\cos\frac{r'}{k}+\sin\frac{r}{k}\sin\frac{r'}{k}\cos(\theta-\theta')$$

$$=zz'+\frac{xx'}{k^2}+\frac{yy'}{k^2},$$

or, in terms of the ratios of the coordinates,

$$\cos\frac{d}{k}=\frac{xx'+yy'+k^2zz'}{\sqrt{x^2+y^2+k^2z^2}\sqrt{x'^2+y'^2+k^2z'^2}}$$

It is convenient to introduce here a brief notation. If $(x, y, z)$, $(x', y', z')$ are the coordinates of two points, we shall define

$$xx' + yy' + k^2zz' \equiv (xx'),$$

and we shall speak of the points $(x)$ and $(x')$.

Then the distance between the points $(x)$ and $(x')$ is given by

$$\cos\frac{d}{k} = \frac{(xx')}{\sqrt{(xx)}\sqrt{(x'x')}}.$$

5. In elliptic geometry the distance-function is periodic.

Suppose $d = \frac{1}{2}\pi k$; then $\cos\frac{d}{k} = 0$, and

$$xx' + yy' + k^2zz' = 0,$$

*i.e.* all points on this line are at the distance $\frac{1}{2}\pi k$ or a quadrant from $(x', y', z')$. This is therefore the equation of the absolute polar of $(x', y', z')$. It is the polar with respect to the conic

$$x^2 + y^2 + k^2z^2 = 0.$$

This is therefore the equation of the absolute.

Suppose $d = \pi k$; then $\cos\frac{d}{k} = -1$, and, with actual values of the coordinates,

$$xx' + yy' + k^2zz' = -k^2,$$

but

$$x^2 + y^2 + k^2z^2 = k^2,$$

and

$$x'^2 + y'^2 + k^2z'^2 = k^2;$$

therefore, multiplying the first equation by 2 and adding to the others,

$$(x + x')^2 + (y + y')^2 + k^2(z + z')^2 = 0,$$

which requires that $x' = -x$, $y' = -y$, $z' = -z$.

In spherical geometry these would represent antipodal points. In elliptic geometry antipodal points coincide, and therefore in every case, if two points have their coordinates in the same ratios, they must coincide.

**6. Angle between two lines.**

From the figure (Fig. 79) we have

$$\sin\frac{p_1}{k}=\sin\frac{r}{k}\sin\phi_1,\quad \cos\beta_1=\cot\frac{r}{k}\tan\frac{p_1}{k},$$

$$\sin\frac{p_2}{k}=\sin\frac{r}{k}\sin\phi_2,\quad \cos\beta_2=\cot\frac{r}{k}\tan\frac{p_2}{k},$$

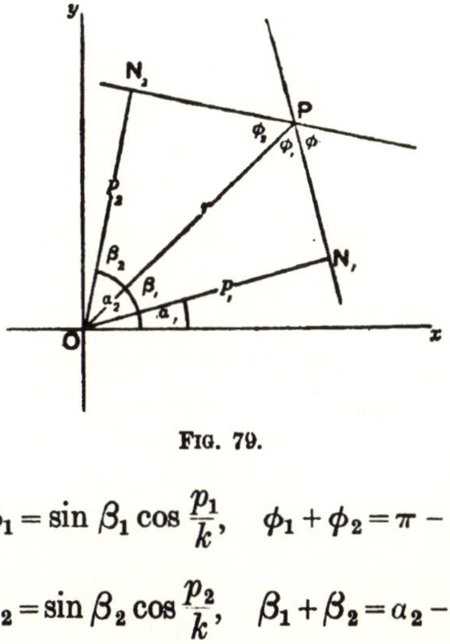


FIG. 79.

$$\cos\phi_1=\sin\beta_1\cos\frac{p_1}{k},\quad \phi_1+\phi_2=\pi-\phi,$$

$$\cos\phi_2=\sin\beta_2\cos\frac{p_2}{k},\quad \beta_1+\beta_2=\alpha_2-\alpha_1,$$

$$\cos(\phi_1+\phi_2)=\sin\beta_1\sin\beta_2\cos\frac{p_1}{k}\cos\frac{p_2}{k}$$

$$-\operatorname{cosec}^2\frac{r}{k}\sin\frac{p_1}{k}\sin\frac{p_2}{k},$$

$$\cos(\beta_1+\beta_2)=\cot^2\frac{r}{k}\tan\frac{p_1}{k}\tan\frac{p_2}{k}-\sin\beta_1\sin\beta_2$$

$$=\cos(\alpha_2-\alpha_1).$$

Therefore

$$\cos\phi = \operatorname{cosec}^2\frac{r}{k}\sin\frac{p_1}{k}\sin\frac{p_2}{k}$$

$$+\left(\cos\overline{a_2 - a_1} - \cot^2\frac{r}{k}\tan\frac{p_1}{k}\tan\frac{p_2}{k}\right)\cos\frac{p_1}{k}\cos\frac{p_2}{k}$$

$$=\sin\frac{p_1}{k}\sin\frac{p_2}{k} + \cos\frac{p_1}{k}\cos\frac{p_2}{k}\cos(a_2 - a_1)$$

$$=\frac{\zeta_1\zeta_2}{k^2} + \xi_1\xi_2 + \eta_1\eta_2,$$

or, in terms of the ratios,

$$\cos\phi = \frac{k^2\xi_1\xi_2 + k^2\eta_1\eta_2 + \zeta_1\zeta_2}{\sqrt{k^2\xi_1^2 + k^2\eta_1^2 + \zeta_1^2}\sqrt{k^2\xi_2^2 + k^2\eta_2^2 + \zeta_2^2}}.$$

If $(\xi\xi) = 0$ is the line-equation of the absolute,

$$\cos\phi = \frac{(\xi_1\xi_2)}{\sqrt{(\xi_1\xi_1)}\sqrt{(\xi_2\xi_2)}}.$$

## 7. Distance of a point from a line.

If $d$ is the distance of a point from a line, $\frac{1}{2}\pi k - d$ is the distance of the point from the pole of the line. Let the coordinates of the point be $(x, y, z)$ and of the line $(\xi, \eta, \zeta)$. The pole of the line is $(k^2\xi, k^2\eta, \zeta)$. Therefore

$$\sin\frac{d}{k} = \frac{\xi x + \eta y + \zeta z}{\sqrt{x^2 + y^2 + k^2z^2}\sqrt{\xi^2 + \eta^2 + \zeta^2/k^2}} = \frac{\xi x + \eta y + \zeta z}{\sqrt{(xx)}\sqrt{(\xi\xi)}}.$$

## 8. Point of intersection of two lines $(\xi_1, \eta_1, \zeta_1)$, $(\xi_2, \eta_2, \zeta_2)$.

The coordinates of the point of intersection are proportional to $(\eta_1\zeta_2 - \eta_2\zeta_1, \zeta_1\xi_2 - \zeta_2\xi_1, \xi_1\eta_2 - \xi_2\eta_1)$.

If the actual values are $\alpha$, $\beta$, $\gamma$, so that $\alpha=\lambda x$, etc., then

$$\alpha^2+\beta^2+k^2\gamma^2=k^2=\lambda^2[(\eta_1\zeta_2-\eta_2\zeta_1)^2+(\zeta_1\xi_2-\zeta_2\xi_1)^2 + k^2(\xi_1\eta_2-\xi_2\eta_1)^2]$$

$$=\lambda^2k^2[(\xi_1^2+\eta_1^2+\zeta_1^2/k^2)(\xi_2^2+\eta_2^2+\zeta_2^2/k^2) - (\xi_1\xi_2+\eta_1\eta_2+\zeta_1\zeta_2/k^2)^2]$$

$$=\lambda^2k^2[1-\cos^2\phi];$$

therefore $$\lambda=\operatorname{cosec}\phi,$$

where $\phi$ is the angle between the lines.

If the coordinates of the point of intersection satisfy the equation $x^2+y^2+k^2z^2=0$, *i.e.* if the lines intersect on the absolute, $\lambda$ is infinite and $\phi$ is zero. The two lines in this case are parallel.

If the ratios of the coordinates make $x^2+y^2+k^2z^2<0$, $\lambda$ is imaginary. The two lines have then no real point of intersection, and the angle $\phi$ is imaginary. The lines may be said to intersect outside the absolute. (These two cases can, of course, only happen in hyperbolic geometry.)

In the latter case the two lines have a common perpendicular.

Let $\xi x+\eta y+\zeta z=0$ be perpendicular to both; then

$$\xi\xi_1+\eta\eta_1+\zeta\zeta_1/k^2=0, \quad \xi\xi_2+\eta\eta_2+\zeta\zeta_2/k^2=0;$$

therefore $\xi:\eta:\zeta=\eta_1\zeta_2-\eta_2\zeta_1:\zeta_1\xi_2-\zeta_2\xi_1:k^2(\xi_1\eta_2-\xi_2\eta_1)$;

but this line is just the polar of their point of intersection. The length $p$ of the common perpendicular is equal to $k\phi$, and we have

$$\cos\phi=\cos\frac{p}{k}=\xi_1\xi_2+\eta_1\eta_2+\frac{\zeta_1\zeta_2}{k^2}.$$

The actual Weierstrass coordinates of an ideal point are therefore purely imaginary numbers of the form $(ix, iy, iz)$, and their ratios are real. If we let the coordinates $(x, y, z)$ be any complex numbers, we get points belonging to the whole "complex domain." This includes (1) real actual points, for which the ratios $x : y : z$ are real and $x^2+y^2+k^2z^2$ has the same sign as $k^2$. (2) real ideal points, for which the ratios $x : y : z$ are real, and $(x^2+y^2+k^2z^2)/k^2$ is negative, (3) imaginary points, for which at least one of the ratios $x : y : z$ is imaginary. The line joining a pair of conjugate imaginary points is a real line, actual, at infinity or ideal. The distance between a pair of conjugate imaginary points is real only if their join is ideal.

## 9. Line joining two points.

Similarly the line-coordinates of the line joining two points $(x_1, y_1, z_1)$, $(x_2, y_2, z_2)$ are proportional to $y_1z_2-y_2z_1$, $z_1x_2-z_2x_1$, $x_1y_2-x_2y_1$. The actual values of the line-coordinates are found by multiplying by the factor cosec $\frac{d}{k}$, where $d$ is the distance between the two points.

If the ratios of the line-coordinates satisfy the equation $\xi^2+\eta^2+\zeta^2/k^2=0$, the line is at infinity, and the distance $d$ is zero.

If the ratios make $\xi^2+\eta^2+\zeta^2/k^2<0$, the line is wholly ideal, and the distance $d$ is imaginary.

## 10. Minimal lines.

When the join of two points is a tangent to the absolute, the distance between the two points is zero. For this reason the tangents to the absolute are called *minimal lines*.

In euclidean geometry the distance between two points $(x_1, y_1)$, $(x_2, y_2)$ is zero if

$$(x_2-x_1)^2+(y_2-y_1)^2=0,$$

*i.e.* if

$$y_2-y_1=\pm i(x_2-x_1),$$

*i.e.* if the join of the two points passes through one of the circular points (Chap. II. § 17). The line at infinity itself passes through both of the circular points, and it is the only real line which passes through them. The distance between two points at infinity should thus be zero. But again, any point on the line at infinity is infinitely distant from any other point. Hence the distance between two points, both of which are at infinity, becomes indeterminate. In relation to the rest of the plane we must consider such distances as infinite, and the geometry of points at infinity becomes quite unmanageable. The geometry upon the line at infinity by itself, however, is really elliptic, since the absolute upon this line consists of a pair of imaginary points; the "distance" between two points at infinity could then be represented by the angle which they subtend at any finite point.

## 11. Concurrency and collinearity.

The condition that the lines $(\xi_1, \eta_1, \zeta_1)$, etc., be concurrent is

$$\begin{vmatrix} \xi_1 & \eta_1 & \zeta_1 \\ \xi_2 & \eta_2 & \zeta_2 \\ \xi_3 & \eta_3 & \zeta_3 \end{vmatrix} = 0.$$

The condition that the points $(x_1, y_1, z_1)$, etc., be collinear is

$$\begin{vmatrix} x_1 & y_1 & z_1 \\ x_2 & y_2 & z_2 \\ x_3 & y_3 & z_3 \end{vmatrix} = 0.$$

These conditions are, of course, the same as those in ordinary analytical geometry, with homogeneous coordinates.

Since the equation of a straight line is homogeneous and of the first degree in the coordinates, all theorems of ordinary geometry which do not involve the actual values of the coordinates, or the distance-formulae, will be true also in non-euclidean geometry. These theorems are those of projective geometry. The difference between euclidean and non-euclidean geometry only appears in the form of the identical relation which connects the point and line coordinates, *i.e.* in the form of the absolute.

## 12. The circle.

A circle is the locus of points equidistant from a fixed point. Let $(x_1, y_1, z_1)$ be the centre and $r$ the radius; then the equation of the circle is

$$\cos\frac{r}{k}=\frac{xx_1+yy_1+zz_1}{\sqrt{x^2+y^2+k^2z^2}\sqrt{x_1{}^2+y_1{}^2+k^2z_1{}^2}},$$

or, when rationalised,

$$(xx)(x_1x_1)\cos^2\frac{r}{k}=(xx_1)^2.$$

This equation is of the second degree, and from its form we see that it represents a conic touching the absolute $(xx)=0$ at the points where it is cut by the line $(xx_1)=0$. $(xx_1)=0$ is the polar of the centre, and is therefore equidistant from the circle, *i.e.* it is the axis of the circle. Hence *A circle is a conic having double contact with the absolute; its axis is the common chord and its centre is the pole of the common chord.*

*The equidistant-curve.* Let $(\xi_1, \eta_1, \zeta_1)$ be the coordinates of the axis, and $d$ the constant distance; then the equation of the curve is

$$\sin\frac{d}{k}=\frac{\xi_1x+\eta_1y+\zeta_1z}{\sqrt{\xi_1{}^2+\eta_1{}^2+\zeta_1{}^2/k^2}\sqrt{x^2+y^2+k^2z^2}},$$

or

$$(xx)(\xi\xi)\sin^2\frac{d}{k}=(\xi_1x+\eta_1y+\zeta_1z)^2.$$

This again represents a conic having double contact with the absolute, the common chord being the axis. The pole of the axis is equidistant from the curve, and so the equidistant-curve is a circle. In elliptic geometry both centre and axis are real, in hyperbolic geometry the centre alone is real for a proper circle, and the axis alone is real for an equidistant-curve.

*The horocycle.* In hyperbolic geometry, the equation of the absolute being $x^2+y^2-k^2z^2=0$, the equation of a horocycle is of the form

$$x^2+y^2-k^2z^2=\lambda(ax+by+cz)^2,$$

where $$a^2+b^2=\frac{c^2}{k^2}.$$

## 13. Coordinates of a point dividing the join of two points into given parts.

If $(x_1, y_1, z_1)$, $(x_2, y_2, z_2)$ are any two points, the coordinates of any point on the line joining them are

$$(\lambda x_1+\mu x_2,\ \ \lambda y_1+\mu y_2,\ \ \lambda z_1+\mu z_2),$$

for if $ax+by+cz=0$ is the equation of the line, so that it is satisfied by the coordinates of the two given points, it will be satisfied also by the coordinates of any point with coordinates of this form. Similarly, if we consider these as the line-coordinates of two lines, the coordinates of any line through their point of intersection are of this form. In fact the line

$$\lambda(a_1x+b_1y+c_1z)+\mu(a_2x+b_2y+c_2z)=0,$$

whose coordinates are $(\lambda a_1+\mu a_2, \ldots)$, passes through the intersection of the two given lines $a_1x+b_1y+c_1z=0$ and $a_2x+b_2y+c_2z=0$.

To find the coordinates of a point dividing the join of two points whose actual coordinates are $(x_1, y_1, z_1)$ and $(x_2, y_2, z_2)$ into two parts $r_1$ and $r_2$, where $r_1+r_2=r$.

Let $(\lambda x_1+\mu x_2, \ldots)$ be the actual coordinates of the required point. Then

$$x_1(\lambda x_1+\mu x_2)+y_1(\lambda y_1+\mu y_2)+k^2z_1(\lambda z_1+\mu z_2)=k^2\cos\frac{r_1}{k};$$

therefore $$\lambda+\mu\cos\frac{r}{k}=\cos\frac{r_1}{k}.$$

Similarly $$\lambda \cos \frac{r}{k} + \mu = \cos \frac{r_2}{k};$$

whence $\lambda \sin \frac{r}{k} = \sin \frac{r_2}{k}$ and $\mu \sin \frac{r}{k} = \sin \frac{r_1}{k},$

and the actual coordinates are

$$\frac{x_1 \sin \frac{r_2}{k} + x_2 \sin \frac{r_1}{k}}{\sin \frac{r}{k}}, \ldots .$$

If $(x_1, y_1, z_1)$, etc., are only the ratios of the coordinates, we must first find their actual values by dividing by the factor $(xx)/k$.

If the line is divided externally into two parts $r_1$ and $r_2$, we have only to observe the proper signs of $r$, $r_1$ and $r_2$.

**14. Middle point of a segment.**

In particular, if $r_1 = r_2$ we get the ratios of the coordinates of the middle point of the segment $(x_1 + x_2, y_1 + y_2, z_1 + z_2)$, or, if $x_1$, etc., are only proportional to the coordinates, the ratios of the coordinates of the middle point are

$$\frac{x_1}{\sqrt{(x_1x_1)}} + \frac{x_2}{\sqrt{(x_2x_2)}} : \frac{y_1}{\sqrt{(x_1x_1)}} + \frac{y_2}{\sqrt{(x_2x_2)}} : \frac{z_1}{\sqrt{(x_1x_1)}} + \frac{z_2}{\sqrt{(x_2x_2)}},$$

the actual values being obtained by dividing by $2 \cos \frac{r}{2k}$.

The join of two points has a second middle point with coordinates $\frac{x_1}{\sqrt{(x_1x_1)}} - \frac{x_2}{\sqrt{(x_2x_2)}} : . : .$, the actual values being obtained by dividing by $2 \sin \frac{r}{2k}$. In elliptic geometry these points are both real and a quadrant apart;[1] in

[1] In spherical geometry the two middle points of a segment are antipodal, and are not (as in elliptic geometry) harmonic conjugates with respect to the given points.

hyperbolic geometry the factor $2\sin\frac{r}{2k}$ becomes $2i\sinh\frac{r}{2k}$, and the coordinates of the second middle point are all imaginary.

**15. Properties of triangles. Centroid, in- and circum-centres.**

Through each vertex of a triangle $(x_1)$, $(x_2)$, $(x_3)$ pass two *medians*, and the medians are concurrent in sets of three in four *centroids*, denoted, in the notation of § 4, by

$$\left(\frac{x_1}{\sqrt{(x_1x_1)}} \pm \frac{x_2}{\sqrt{(x_2x_2)}} \pm \frac{x_3}{\sqrt{(x_3x_3)}}\right).$$

The same combination of signs is taken for all three coordinates, and there are four different combinations of signs, one corresponding to each of the centroids.

Similarly, the middle points of the sides are collinear in sets of three in four lines, the axes of the circumscribed circles.

The bisectors of the angles are concurrent in sets of three in four points, the centres of the inscribed circles; and their points of intersection with the opposite sides are collinear in sets of three in four lines.

**16. Explanation of apparent exception in euclidean geometry.**

In euclidean geometry four circles can be drawn to touch the sides of a triangle, but apparently only one can be circumscribed. Of the four circumcircles of a triangle in hyperbolic geometry, three are equidistant-curves. In euclidean geometry the equidistant-curve through $B$, $C$ and $A$ reduces to the line $BC$ and the line through $A \parallel BC$. (Cf. Chap. II. § 23.)

The conception of a pair of parallel straight lines as forming a circle in euclidean geometry is consistent with the definition of a circle as a conic having double contact with the absolute, for the absolute in this case is a pair of coincident straight lines, and this is cut by a pair of parallel lines in two pairs of coincident points. A single straight line is not, of course, a tangent to the absolute, though it cuts it in two coincident points; this case is just the same as that of a line which passes through a double point on a curve, but which is not considered as being a tangent. But when we have a pair of parallel lines cutting the absolute $\Omega$ in four points all coincident, we can regard $\Omega$ as being a tangent to the curve consisting of this pair of lines. Fig. 80 represents the case approximately when the absolute is still a proper conic and the pair of straight lines is also a proper conic, having double contact with the absolute.

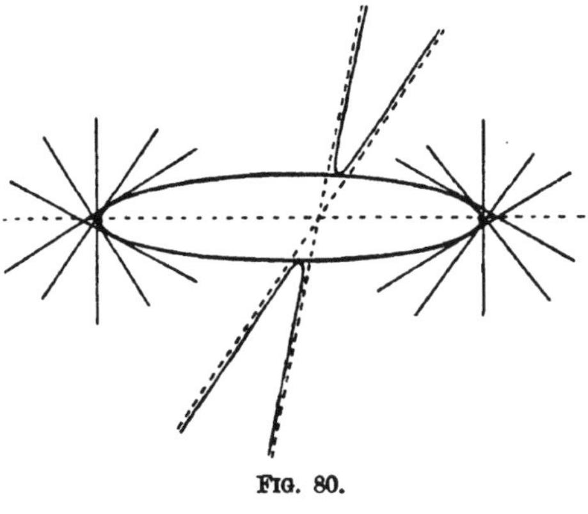

FIG. 80.

The axis of the circle consisting of a pair of parallel lines is the line lying midway between them; the absolute pole of this (a point at infinity) is the centre. When the axis passes through the centre, *i.e.* when it coincides with the line at infinity, the circle becomes a horocycle, which is thus represented in euclidean geometry by a straight line together with the line at infinity.

Two equidistant-curves, with parallel axes, have the same centre at infinity. In hyperbolic geometry two equidistant-curves, with parallel axes intersecting at infinity at $O$, have their centres on the tangent at $O$, and therefore at a zero distance apart though not coincident.

## 17. Polar triangles. Orthocentre and orthaxis.

If $A$, $B$, $C$ is a triangle and $A'$, $B'$, $C'$ the absolute poles of the sides $a$, $b$, $c$, then the sides $a'$, $b'$, $c'$ of the second triangle are the absolute polars of the vertices $A$, $B$, $C$ of the given triangle. Two such triangles are called *polar triangles*.

If the coordinates of $A$, $B$, $C$ are $(x_1, y_1, z_1)$, etc., the equations of their polars are $(xx_1)=0$, etc.

The point-coordinates of the vertices $A'$, $B'$, $C'$ are $y_1z_2-y_2z_1,\ z_1x_2-z_2x_1,\ (x_1y_2-x_2y_1)/k^2$, etc.

The equation of $AA'$, which joins $(x_1, y_1, z_1)$ to the point of intersection of $(xx_2)=0$ and $(xx_3)=0$, is

$$(xx_2)(x_3x_1)-(xx_3)(x_1x_2)=0.$$

Writing down two other equations by a cyclic permutation of the suffixes, we get the equations of $BB'$ and $CC'$, and the sum of these vanishes identically. Hence $AA'$, $BB'$, $CC'$ are concurrent. $AA' \perp BC$ and $B'C'$; hence the point of concurrence is the common *orthocentre* $O$ of the triangles $ABC$, $A'B'C'$.

The absolute poles of $AA'$, $BB'$, $CC'$, *i.e.* the points on the sides of the triangles distant a quadrant from the opposite vertices, will be collinear in a line called the *orthaxis*, *o*, which is the absolute polar of the orthocentre.

The two triangles $ABC$, $A'B'C'$ are *in perspective* with centre $O$ and axis $o$.

## 18. Desargues' theorem. Configurations.

The last result is a particular case of Desargues' theorem for perspective triangles, which, since it expresses a projective property, is true in non-euclidean geometry, and can be proved (using space of three dimensions) in a purely projective manner.

In the figure for Desargues' theorem (Fig. 81) we have two triangles with their corresponding vertices lying on

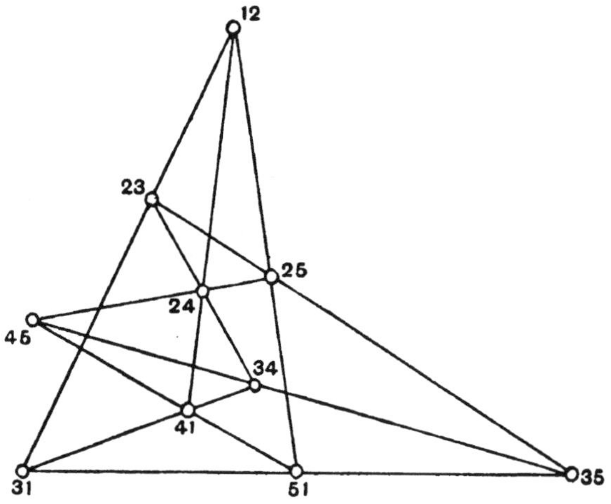


FIG. 81.

three concurrent lines, and their corresponding sides intersecting in three collinear points. There are thus 10 points and 10 lines: through each point pass 3 lines, and on each line lie 3 points. A figure of points and lines with this property, that through every point pass the same number of lines and on every line lie the same number of points, is called a *configuration*. If $p_{01}$ denotes the number of lines through a point, $p_{10}$ the number of points on a line, $p_{00}$ the whole number of points, and $p_{11}$ the whole number of lines, the configuration may be denoted by the symbol

$$\begin{array}{|cc|} p_{00} & p_{01} \\ p_{10} & p_{11} \end{array}\,.$$

Desargues' configuration is represented by

$$\begin{array}{|cc|} 10 & 3 \\ 3 & 10 \end{array}\,,$$

and is *reciprocal*. A convenient notation for the points is by pairs of the numbers from 1 to 5. The three points which lie on one line are denoted by the combinations with the same three numbers.

The configuration formed by the six middle points $M_{01}$, $M_{23}$, etc., of the sides of a triangle $ABC$ and the four points of concurrency $G_0$, $G_1$, $G_2$, $G_3$ of the medians is a Desargues configuration of a special kind (Fig. 82). The points $G$ form a complete quadrangle, and the points $M$ are the vertices of a complete quadrilateral, both having $ABC$ as diagonal triangle. This is called, therefore, the *quadrangle-quadrilateral configuration*. Each vertex $M_{rs}$ of the quadrilateral lies on a side $G_rG_s$ of the quadrangle.

Similarly, the six bisectors of the angles and the four lines of collinearity of the points in which they meet the sides of the triangle form the same configuration.

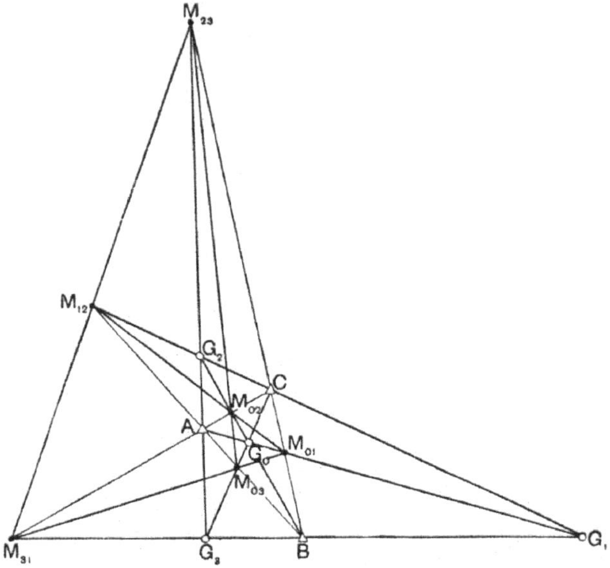


FIG. 82.

**19. Desmic system.**

In three dimensions we have similar interesting configurations.

If $(x_1)$, $(x_2)$, $(x_3)$, $(x_4)$ are four points in space,

$$\left(\frac{x_1}{\sqrt{(x_1x_1)}} \pm \frac{x_2}{\sqrt{(x_2x_2)}} \pm \frac{x_3}{\sqrt{(x_3x_3)}} \pm \frac{x_4}{\sqrt{(x_4x_4)}}\right)$$

represent the eight centroids of the four points. Each centroid is on a line joining one of the points to the centroid of the other three.

If the four given points be denoted by $A_1$, $A_2$, $A_3$, $A_4$, and the other points corresponding to the different combinations of signs be represented as follows :

$$\begin{array}{lllll} + & + & + & + & B_1, \\ + & + & - & - & B_2, \\ + & - & + & - & B_3, \\ + & - & - & + & B_4, \end{array} \qquad \begin{array}{lllll} + & - & - & - & C_1, \\ + & - & + & + & C_2, \\ + & + & - & + & C_3, \\ + & + & + & - & C_4, \end{array}$$

then the join of any $B$ with any $C$ passes through an $A$, *e.g.* $B_2 - C_4$ gives $A_3$. So the 12 points lie in sets of 3 on 16 lines. They form three tetrahedra, any two of which are in perspective in four different ways, the centres of perspective being the vertices of the third tetrahedron. Corresponding planes of two perspective tetrahedra intersect in four lines which are coplanar, and these planes are the faces of the third tetrahedron. A system of tetrahedra of this kind is called a *desmic system.*

In a similar way it may be proved that the centres or axial planes of the 8 circum- or in-scribed spheres form with the given tetrahedron a desmic system.

A simple example of a desmic system in ordinary space is afforded by the corners of a cube, its centre and the points of concurrency (at infinity) of its edges.

## 20. Concurrency and collinearity.

In euclidean geometry we have the two useful theorems of Menelaus and Ceva as tests for collinearity and concurrency. Theorems corresponding to these hold also in non-euclidean geometry.

I. If a transversal meets the sides of a triangle $ABC$ in $XYZ$, and $\alpha$, $\beta$, $\gamma$ are the angles of intersection, taken positively, we have (Fig. 83)

$$\frac{\sin BX}{\sin BZ} = -\frac{\sin\gamma}{\sin\alpha}, \quad \frac{\sin CY}{\sin CX} = \frac{\sin\alpha}{\sin\beta}, \quad \frac{\sin AZ}{\sin AY} = -\frac{\sin\beta}{\sin\gamma},$$

the positive directions on the sides being in the cyclic order $ABC$. Hence

$$\frac{\sin BX}{\sin CX}\cdot\frac{\sin CY}{\sin AY}\cdot\frac{\sin AZ}{\sin BZ}=+1.$$

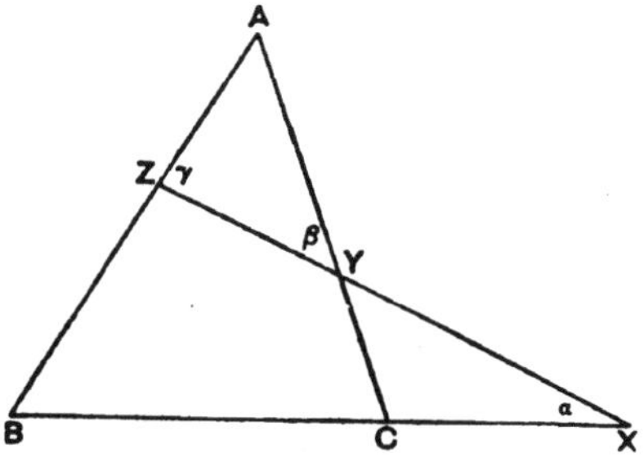


FIG. 83.

II. If three concurrent lines through the vertices meet the opposite sides of a triangle $ABC$ in $XYZ$, and $\alpha$, $\beta$, $\gamma$ are the angles between the lines (Fig. 84),

$$\frac{\sin BX}{\sin OB}=\frac{\sin\gamma}{\sin X}\quad\text{and}\quad\frac{\sin CX}{\sin OC}=\frac{\sin\beta}{\sin X};$$

therefore $$\frac{\sin BX}{\sin CX}=-\frac{\sin OB}{\sin OC}\cdot\frac{\sin\gamma}{\sin\beta}.$$

Similarly $$\frac{\sin CY}{\sin AY}=-\frac{\sin OC}{\sin OA}\cdot\frac{\sin\alpha}{\sin\gamma},$$

and $$\frac{\sin AZ}{\sin BZ}=-\frac{\sin OA}{\sin OB}\cdot\frac{\sin\beta}{\sin\alpha}.$$

Therefore $$\frac{\sin BX}{\sin CX}\cdot\frac{\sin CY}{\sin AY}\cdot\frac{\sin AZ}{\sin BZ}=-1.$$

Conversely, the points $X$, $Y$, $Z$ are collinear, or $AX$, $BY$, $CZ$ are concurrent, according as

$$\frac{\sin BX}{\sin CX}\cdot\frac{\sin CY}{\sin AY}\cdot\frac{\sin AZ}{\sin BZ}=+1\text{ or }-1.$$

This condition may be put in another form. Since $\dfrac{\sin BX}{\sin CX}=\dfrac{\sin AB}{\sin AC}\cdot\dfrac{\sin BAX}{\sin CAX}$, the condition reduces to

$$\frac{\sin BAX}{\sin CAX}\cdot\frac{\sin CBY}{\sin ABY}\cdot\frac{\sin ACZ}{\sin BCZ}=\pm 1,$$

in which form it is the same as the condition in euclidean geometry.

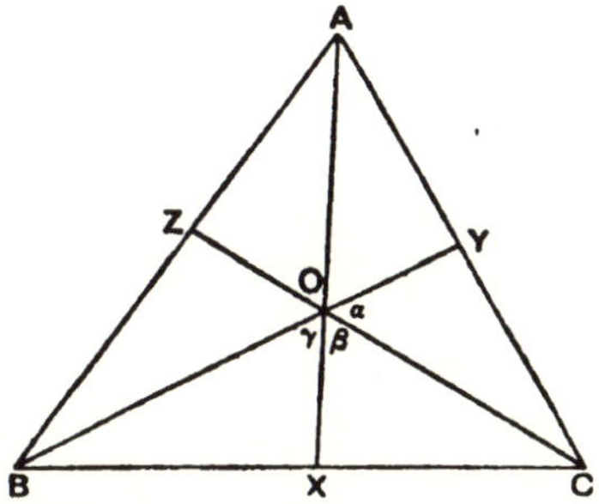


FIG. 84.

From this it follows at once that if $AX$, $BY$, $CZ$ are three concurrent lines through $O$, their *isogonal conjugates* with respect to the sides of the triangle are concurrent in the isogonal conjugate of $O$.

## 21. Position-ratio. Cross-ratio.

If $X$, $Y$, $P$ are collinear, the ratio $\dfrac{\sin XP}{\sin YP}$ is called the *position-ratio* of $P$ with respect to $X$ and $Y$, and the double ratio $\dfrac{\sin XP}{\sin YP}\div\dfrac{\sin XQ}{\sin YQ}$ is called the *cross-ratio* of the range $(XY, PQ)$.

Similar definitions can be given for pencils of rays, and the whole theory of cross-ratio can be developed on the same lines as in ordinary geometry.

Thus, the cross-ratio of a pencil is equal to that of any transversal, and cross-ratios are unaltered by projection.

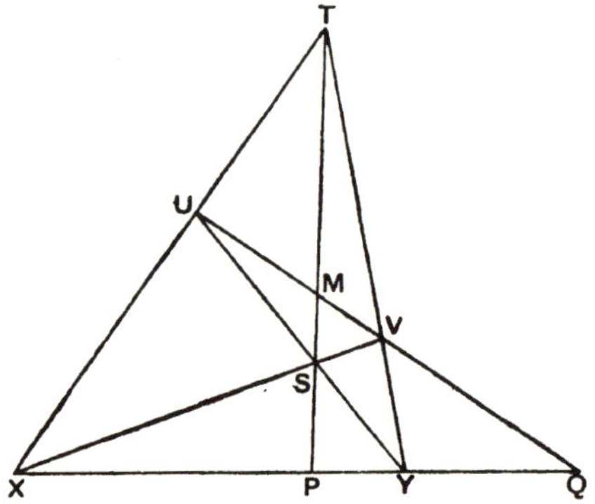


FIG. 85.

Further, it can be shown that

$$(ABCD)=(BADC)=(CDAB)=(DCBA),$$

$$(ABCD)\,.\,(ABDC)=1, \quad (ABCD)+(ACBD)=1.$$

The harmonic property of the complete quadrilateral follows.

For (Fig. 85), $(XYPQ)\,\overline{\wedge}_T(UVMQ)$, and also $\overline{\wedge}_S(VUMQ)$.

Therefore $\qquad (UV, MQ)=(VU, MQ)=-1.$

If $A(x_1, y_1, z_1)$ and $B(x_2, y_2, z_2)$ are two fixed points, and $P$ a variable point with coordinates

$$(x_1+\lambda x_2, \quad y_1+\lambda y_2, \quad z_1+\lambda z_2),$$

then, if $\qquad AP=r_1, \quad PB=r_2, \quad AB=r,$

we found $\lambda=\sin\frac{r_1}{k}\Big/\sin\frac{r_2}{k}=$ the position-ratio of $P$ with respect to $A$ and $B$. If $Q$ is the point corresponding to the parameter $\mu$, the cross-ratio $(AB, PQ)=\frac{\lambda}{\mu}$. The cross-ratio of the two pairs of points corresponding to the parameters $\lambda$, $\lambda'$ and $\mu$, $\mu'$ is

$$(\lambda\lambda',\ \mu\mu') = \frac{\lambda - \mu}{\lambda - \mu'} \div \frac{\lambda' - \mu}{\lambda' - \mu'}.$$

These results are the same as in euclidean geometry.

### EXAMPLES IV.

**1.** Prove that the actual Weierstrass line-coordinates of the absolute polar of $(x, y, z)$ are $(x/k, y/k, kz)$, and the actual point-coordinates of the absolute pole of $(\xi, \eta, \zeta)$ are $(k\xi, k\eta, \zeta/k)$.

**2.** If the distance between the points $(x_1, y_1, z_1)$, $(x_2, y_2, z_2)$ vanishes, prove that their join touches the absolute.

**3.** If $(x_1 + ix_2, y_1 + iy_2, z_1 + iz_2)$ are the actual Weierstrass coordinates of a point ($x_1$, $y_1$, etc., being real numbers), prove that $(x_1, y_1, z_1)$ and $(x_2, y_2, z_2)$ are conjugate with regard to the absolute.

**4.** If $(x_1 + ix_2, \ldots)$ $(a_1 + ia_2, \ldots)$ are the actual Weierstrass coordinates of two points at a real distance ($x_1$, $y_1$, etc., being real numbers), prove that, for all values of $\lambda$, $(x_1 + \lambda a_1, \ldots)$ and $(x_2 + \lambda a_2, \ldots)$ are conjugate with regard to the absolute.

**5.** If $ds$ is the element of arc of a curve and $dx$, $dy$, $dz$ the differentials of the Weierstrass coordinates, prove that $ds^2 = dx^2 + dy^2 + k^2dz^2$.

If $r$, $\theta$ are the polar coordinates, prove that $ds^2 = dr^2 + k^2 \sin^2\frac{r}{k}\,d\theta^2$.

**6.** $ABCD$ is a skew quadrilateral, $PQRS$ are points on the four sides $AB$, $BC$, $CD$, $DA$. Prove that if

$$\sin AP \sin BQ \sin CR \sin DS = \sin BP \sin CQ \sin DR \sin AS,$$

the four points $PQRS$ lie in one plane.

**7.** 1, 2, 3, 4 are the vertices of a tetrahedron. A plane cuts each of the six edges. If the edge 12 is cut at $A$, and the ratio $\sin 1A/\sin 2A$ is denoted by (12), prove that $(12)(23)(34)(41) = 1$. Conversely, if $(12)(23)(34)(41) = 1$, prove that the points 12, 23, 34, 41 (*i.e.* the corresponding points on these edges) are coplanar.

**8.** If $(12)(23)(34)(41) = 1 = (12)(24)(43)(31) = (13)(32)(24)(41)$, prove that either (i) the sets of points 12, 23, 31, etc., are collinear, or (ii) the lines (12, 34), (13, 24), (14, 23) are concurrent.

**9.** Four circles touch in succession, each one touching two others (the number of external contacts being even); show that the four points of contact lie on a circle, and that the four tangents at the points of contact touch a circle.

**10.** Four spheres touch in succession, each one touching two others (the number of external contacts being even); show that the four points of contact lie on a circle, and that the four tangent planes at the points of contact touch a sphere. Show further that, whatever the nature of the contacts, the four tangent planes pass through one point.

**11.** Five spheres touch in succession, each one touching two others (the number of external contacts being even); show that the five points of contact lie on a sphere, and that the five tangent planes at the points of contact touch a sphere. (*Educ. Times* (n.s.), xi. p. 57.)

**12.** $D$, $E$, $F$ are the feet of the perpendiculars from a point $O$ on the sides of the triangle $ABC$. Prove that

$$\cos BD \cos CE \cos AF = \cos CD \cos AE \cos BF.$$

**13.** $ABC$ is a given triangle, and $l$ is any line. $P$, $Q$, $R$ are the feet of the perpendiculars from $A$, $B$, $C$ on $l$. $PP' \perp BC$, $QQ' \perp CA$, $RR' \perp AB$. Prove that $PP'$, $QQ'$, $RR'$ meet in a point (the *orthopole* of $l$).

**14.** Prove that the locus of a point such that the ratio of the cosines of its distances from two fixed points is constant is a straight line.

**15.** If $L, M, N$; $L_1, M_1, N_1$; etc., are the points of contact of the in- and e-scribed circles of the triangle $ABC$ with the sides $a$, $b$, $c$, and $2s = a+b+c$, prove the relations:

$$AM_1 = AN_1 = BN_2 = BL_2 = CL_3 = CM_3 = s,$$
$$AM = AN = BN_3 = BL_3 = CL_2 = CM_2 = s - a, \text{ etc.}$$

**16.** Establish the reciprocal relations to those in Question 15 for the circumcircles.

**17.** Prove that the envelope of a line which makes with two fixed lines a triangle of constant perimeter is a circle. Prove also that the envelope is a circle if the excess of the sum of two sides over the third side is constant. What is the reciprocal theorem?

(In the following questions, 18-22, the geometry is hyperbolic. The formulae are analogous to well-known formulae in spherical trigonometry.)

**18.** If $ka$, $kb$, $kc$ are the sides, and $A$, $B$, $C$ the angles of a triangle, prove that

$$\cos\frac{A}{2}=\sqrt{\frac{\sinh s\sinh(s-a)}{\sinh b\sinh c}},\quad \sin\frac{A}{2}=\sqrt{\frac{\sinh(s-b)\sinh(s-c)}{\sinh b\sinh c}}.$$

**19.** If $r, r_1, r_2, r_3$ are the radii of the in- and e-scribed circles of a triangle $ABC$, prove that

$$\tanh r\sinh s=\tanh r_1\sinh(s-a)=\tanh r_2\sinh(s-b)$$
$$=\tanh r_3\sinh(s-c)$$
$$=\sqrt{\sinh(s-a)\sinh(s-b)\sinh(s-c)}.$$

**20.** Prove that

$$\tanh r_1\tanh r_2\tanh r_3=\sqrt{\sinh(s-a)\sinh(s-b)\sinh(s-c)}.$$

**21.** If $R$ is the radius of the circumcircle of the triangle $ABC$, prove that

$$2\cosh\frac{a}{2}\cosh\frac{b}{2}\cosh\frac{c}{2}\tanh R=\frac{\sinh a}{\sin A}=\frac{\sinh b}{\sin B}=\frac{\sinh c}{\sin C}.$$

If $D_1$, $D_2$, $D_3$ are the distances of the circumscribed equidistant-curves, prove that

$$2\cosh\frac{a}{2}\sinh\frac{b}{2}\sinh\frac{c}{2}\coth D_1=\frac{\sinh a}{\sin A},\text{ etc.}$$

**22.** Prove that

$$\coth R+\tanh D_1+\tanh D_2+\tanh D_3=2\cosh s\sin A/\sinh a,$$
$$\coth R+\tanh D_1-\tanh D_2-\tanh D_3=2\cosh(s-a)\sin A/\sinh a,\text{ etc.}$$

**23.** Prove that, in the desmic configuration in § 19, the following sets of points are coplanar: $A_1A_2B_1B_2C_1C_2$, $A_3A_4B_3B_4C_1C_2$, and those obtained from these by cyclic permutation of $ABC$ or of 234. Deduce that the configuration has the symbol

| | | |
|---|---|---|
| 12 | 4 | 6 |
| 3 | 16 | 3 |
| 6 | 4 | 12 |

**24.** If one pair of altitudes of a tetrahedron $ABCD$ intersect, prove that the other pair will also intersect; and if one altitude intersects two others, all four are concurrent. If these conditions are satisfied, prove that

$$\cos AB \cos CD = \cos AC \cos BD = \cos AD \cos BC.$$

**25.** Prove that there is a circle which touches the in- and the e-scribed circles of a triangle. [In spherical geometry this is Hart's circle, and corresponds to the nine-point circle in ordinary geometry. See M'Clelland and Preston's *Spherical Trigonometry*, Chap. VI. Art. 88.]

**26.** Prove that there is a circle which touches the four circumcircles of a triangle. [In euclidean geometry the circumscribed equidistant-curves are three pairs of parallel lines and form a triangle $A'B'C'$, of which $A$, $B$, $C$ are the middle points of the sides. The circumcircle of $ABC$ is the nine-point circle of $A'B'C'$, and touches the inscribed circle of $A'B'C'$. That is, the last-named circle touches the four "circumcircles" of the triangle $ABC$.]

## CHAPTER V.

### REPRESENTATIONS OF NON-EUCLIDEAN GEOMETRY IN EUCLIDEAN SPACE.

1. The problem of Representation is one that faces us whenever we try to realise the figures of non-euclidean geometry. There already exists in the mind, whether intuitively or as the result of experience, a more or less clear idea of euclidean geometry. This geometry has from time immemorial been applied to the space in which we live; and now, when it is pointed out to us that there are other conceivable systems of geometry, each as self-consistent as Euclid's, it is a matter of the greatest difficulty to conjure up a picture of space endowed with non-euclidean properties. The image of euclidean space constantly presents itself and suggests as the easiest solution of the difficulty a representation of non-euclidean geometry by the figures of euclidean geometry. Thus, upon a sheet of paper, which is for us the rough model of a euclidean plane, we draw figures to represent the entities of non-euclidean geometry. Sometimes we represent the non-euclidean straight lines by straight lines and sometimes by curves, according as the idea of straightness or that of shape happens to be uppermost in the mind. But we must never forget that the figures that we are constructing are only representations, and that the non-euclidean straight line is

every bit as straight as its euclidean counterpart. Th problem of representing non-euclidean geometry on the euclidean plane is exactly analogous to that of map-projection.

## Projective Representation.

**2.** The fact that a straight line can be represented by an equation of the first degree enables us to represent non-euclidean straight lines by straight lines on the euclidean plane. Distances and angles will not, however, be truly [1] represented, and we must find the functions of the euclidean distances and angles which give the actual distances and angles of non-euclidean geometry.

**3.** The absolute is represented by a conic. In hyperbolic geometry this conic is real, in elliptic geometry it is wholly imaginary, but in every case the polar of a real point is a real line. The conic always has a real equation. In the case in which the absolute is a real conic, we could, if we like, represent it by a circle, but except in special cases this does not give any gain in simplicity.

Two lines whose point of intersection is on the absolute are parallel; two lines whose point of intersection lies outside the absolute are non-intersectors. The points outside the absolute have to be regarded as ultra-infinite, and are called ideal points. They are distinguished from other imaginary points by the fact that, while their actual coordinates are all imaginary, the ratios of their coordinates are real. In the present representation they are repre-

[1] In the sense of map-projections; *i.e.* angles which are equal in the euclidean representation, when measured by euclidean standards, do not in general represent equal angles in the non-euclidean geometry, but, again in the sense of map-projections, figures are distorted.

sented by real points; other imaginary points are represented by imaginary points. (Cf. Chap. IV. § 8.)

A real line has two points at infinity, and part of the line lies in the ideal region. A line which touches the absolute

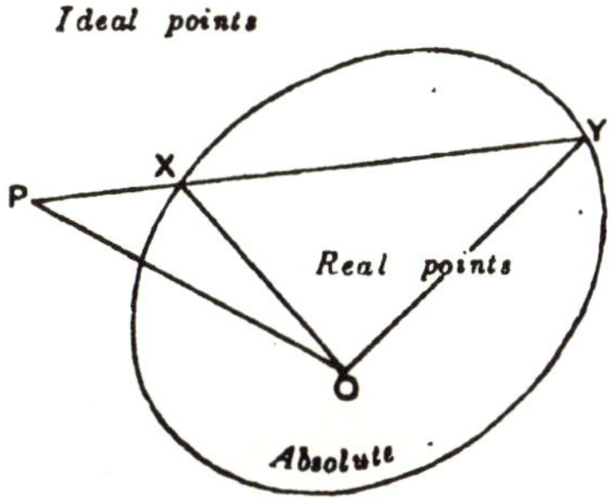


FIG. 86.

has one point at infinity, and all the rest of the line is ideal. A line which lies outside the absolute is wholly ideal.

Through any point two parallels can be drawn to a given line, viz. the points joining the given point to the two points at infinity on the given line. A triangle which has its three vertices on the absolute has a constant area.

In elliptic geometry the absolute is imaginary, and there are no ideal points.

## 4. Euclidean geometry.

Euclidean geometry is a limiting case, where the space-constant $k \to \infty$. The coordinates of a point become the usual rectangular coordinates $x$ and $y$ with $z=1$. The equation of the absolute becomes in point-coordinates $z=0$, and in line-coordinates $\xi^2+\eta^2=0$, *i.e.* the absolute degenerates as a locus to a straight line counted twice—the straight line at infinity, and as an envelope to two imaginary pencils of lines, $\xi+i\eta=0$ and $\xi-i\eta=0$, whose vertices lie on the line at infinity since the line-coordinates of their join are $\xi=0$, $\eta=0$, $\zeta=\zeta$, and its equation is therefore $z=0$. The equations of the lines of these imaginary pencils are of the forms $x+iy+cz=0$, $x-iy+cz=0$.

The formula for the distance between two points,

$$\cos\frac{r}{k}=\frac{xx'+yy'+k^2zz'}{\sqrt{x^2+y^2+k^2z^2}\sqrt{x'^2+y'^2+k^2z'^2}},$$

becomes
$$1-\frac{1}{2}\cdot\frac{r^2}{k^2}=(xx'+yy'+k^2)\cdot\frac{1}{k^2}\cdot\left(1-\frac{1}{2}\cdot\frac{x^2+y^2}{k^2}\right)\left(1-\frac{1}{2}\cdot\frac{x'^2+y'^2}{k^2}\right)$$
$$=\left(1+\frac{xx'+yy'}{k^2}\right)\left(1-\frac{1}{2}\cdot\frac{x^2+y^2+x'^2+y'^2}{k^2}\right)$$
$$=1-\frac{1}{2}\cdot\frac{(x-x')^2+(y-y')^2}{k^2},$$

or
$$r^2=(x-x')^2+(y-y')^2.$$

## 5. The circular points.

The equation of a circle becomes of the general form

$$x^2+y^2+z(ax+by+cz)=0,$$

and this represents a conic passing through the points of intersection of the line $z=0$ with the pair of imaginary lines $x+iy=0$ and $x-iy=0$, *i.e.* every circle passes through the vertices of the imaginary pencils. For this reason these two points are called the *circular points*. This property of the circle is the equivalent of the property that it has double contact with the absolute. (Chap. IV. § 16.)

**6**. Now, in ordinary geometry the angle between two lines can be expressed in terms of the two lines joining their point of intersection to the circular points.[1]

[1] E. Laguerre, "Note sur la théorie des foyers," *Nouv. Ann. Math.*, Paris, **12** (1853).

Let the equations of the two lines $u$, $u'$ through the origin be $y=x\tan\theta$, $y=x\tan\theta'$, and denote the two lines joining $O$ to the circular points by $\omega$, $\omega'$; their equations are $y=ix$, $y=-ix$. The cross-ratio of the pencil $(uu', \omega\omega')$ is

$$\frac{\tan\theta-i}{\tan\theta+i}\div\frac{\tan\theta'-i}{\tan\theta'+i}.$$

Now $$\frac{i-\tan\theta}{i+\tan\theta}=\frac{i\cos\theta-\sin\theta}{i\cos\theta+\sin\theta}=\frac{\cos\theta+i\sin\theta}{\cos\theta-i\sin\theta}=e^{2i\theta}.$$

Therefore $$(uu', \omega\omega')=e^{2i(\theta-\theta')},$$

and $$\phi=\theta'-\theta=\frac{i}{2}\log(uu', \omega\omega'),$$

*i.e. the angle between two lines is a certain multiple of the logarithm of the cross-ratio of the pencil formed by the two lines and the lines joining their point of intersection to the circular points.*

**7.** Now let us return to the case where the absolute is a real conic $x^2+y^2-k^2z^2=0$. Consider two points $P(x, y, z)$, $P'(x', y', z')$. The point $(x+\lambda x', y+\lambda y', z+\lambda z')$ lies on their join. If this point is on the absolute,

$$(x+\lambda x')^2+(y+\lambda y')^2-k^2(z+\lambda z')^2=0,$$

*i.e.* $$\lambda^2(x'^2+y'^2-k^2z'^2)+2\lambda(xx'+yy'-k^2zz')$$
$$+(x^2+y^2-k^2z^2)=0.$$

Let $\lambda_1$, $\lambda_2$ be the roots of this quadratic. The line $PP'$ cuts the absolute in the two points $X$, $Y$, corresponding to these parameters, and the cross-ratio of the range

$$(PP', XY)=\frac{\lambda_1}{\lambda_2}.$$

Let $(PP')=d=k\phi$, and

$$x^2+y^2-k^2z^2=r^2, \quad x'^2+y'^2-k^2z'^2=r'^2;$$

then the quadratic for $\lambda$ becomes

$$\lambda^2 r'^2 + 2\lambda r r' \cos\phi + r^2 = 0;$$

whence $\lambda_1, \lambda_2 = (-\cos\phi \pm \sqrt{-\sin^2\phi}) r/r' = -e^{\pm i\phi} r/r'$.

Therefore $\lambda_1/\lambda_2 = e^{-2i\phi}$ and $\phi = \frac{1}{2} i \log (PP', XY)$.

Therefore $d = \frac{1}{2} ik \log (PP', XY)$,

*i.e. the distance between two points is a certain multiple of the logarithm of the cross-ratio of the range formed by the two points and the two points in which their join cuts the absolute.*

In a similar way it can be shown that *the angle between two straight lines is a certain multiple of the logarithm of the cross-ratio of the pencil formed by the two lines and the two tangents from their point of intersection to the absolute.*

If the unit angle is such that the angle between two lines which are conjugate with regard to the absolute is $\frac{1}{2}\pi$, then

$$\phi = \frac{1}{2} i \log (pp', xy).$$

**8.** By this representation the whole of metrical geometry is reduced to projective geometry, for cross-ratios are unaltered by projection. Any projective transformation which leaves the absolute unaltered will therefore leave distances and angles unaltered. Such transformations are called *congruent transformations* and form the most general motions of rigid bodies.

This projective metric is associated with the name of CAYLEY,[1] who invented the term Absolute. He was the

[1] "A sixth memoir upon quantics," *London Phil. Trans. R. Soc.*, **149** (1859). Cayley wrote a number of papers dealing specially with non-euclidean geometry, but although he must be regarded as one of the epoch-makers, he never quite arrived at a just appreciation of the science. In his mind non-euclidean geometry scarcely attained to an independent existence, but was always either the geometry upon a certain class of curved surfaces, like spherical geometry, or a mode of representation of certain projective relations in euclidean geometry.

first to develop the theory of the absolute, though only as a geometrical representation of the algebra of quantics. KLEIN [1] introduced the logarithmic expressions and showed the connection between Cayley's theory and Lobachevsky's geometry.[2]

**9.** As an example of a projective solution of a metrical problem, let us find the middle points of a segment $PQ$. Let $PQ$ cut the absolute in $X$, $Y$, and let $M_1$, $M_2$ be the double points of the involution $(PQ, XY)$. Then $(XYPM_1) \barwedge (YXQM_1) \barwedge (XYM_1Q)$; therefore dist. $(PM_1)$ = dist. $(M_1Q)$. $M_1$, $M_2$ are therefore the middle points of the segment $(PQ)$.

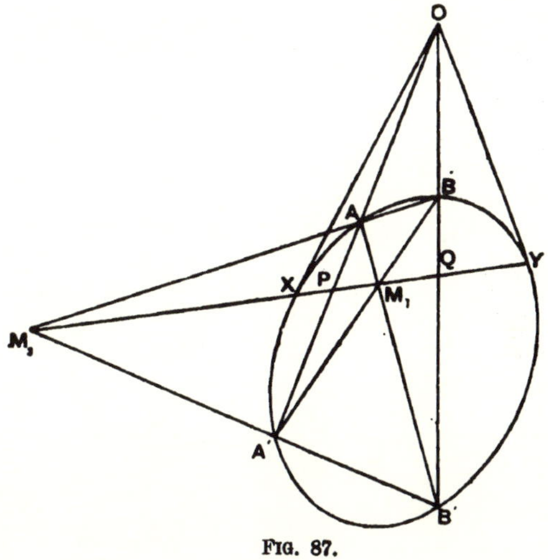


FIG. 87.

Since $M_1$, $M_2$ are harmonic conjugates with respect to $X$, $Y$ and also with respect to $P$, $Q$, the construction is therefore as follows.

[1] "Über die sogenannte Nicht-Euklidische Geometrie," *Math. Ann.*, **4** (1871), **6** (1873).

[2] Since the definition of the cross-ratio of a range is the same in non-euclidean geometry, the logarithmic expressions for distance and angle hold not only in the euclidean representation of the geometry, but also in the actual non-euclidean geometry itself.

Join $O$, the pole of $PQ$, to $P$ and $Q$, cutting the absolute in $AA'$, $BB'$. $AB'$, $A'B$ intersect in $M_1$ and $AB$, $A'B'$ in $M_2$. For by this construction $OM_1M_2$ is a self-conjugate triangle and $M_1$, $M_2$ are harmonic conjugates with respect to $X$, $Y$, and also with respect to $P$, $Q$.

## 10. Classification of geometries with projective metric.

Having arrived at the result that metrical plane geometry is projective geometry in relation to an absolute conic, distances and angles being determined by the projective expressions

$$\text{dist.}\ (PQ) = K \log (XY, PQ), \quad \text{angle}\ (pq) = k \log (xy, pq),$$

we may reverse the process, and define distances and angles by these expressions. We thus get a general system of geometry which will include euclidean, hyperbolic and elliptic geometries as special cases. The nature of the geometry will be determined when the absolute conic is fixed, and the values of the constants $K$ and $k$ have been determined. Generally speaking, the values of these constants depend only on the units of distance and angle which are selected, but there is an essential distinction according as the constants are real, imaginary or infinite. There is no distinction, for example, between the cases corresponding to different real values of $K$. This simply corresponds to a different choice of the arbitrary unit of length; just as in angular measurement the constant $k$ may be chosen so that the measure of a right angle may be $\frac{\pi}{2}$ or 180 or any other number. As each of the constants may conceivably be real, infinite or imaginary, there are nine species of plane geometry.

The points of the absolute are at an infinite distance

from all other points, and the tangents to the absolute make an infinite angle with all other lines.

If the measure of angle is to be the same as in ordinary geometry, the tangents from a real point to the absolute must be imaginary; the cross-ratio $(pq, xy)$ will be imaginary, and $k$ must be purely imaginary. When $p$, $q$ are conjugate with regard to the absolute they are at right angles, and if the unit angle is such that the angle in this case is $\frac{\pi}{2}$, then $k = \frac{1}{2}i$.

Then there are three cases according as the absolute is a real proper conic (hyperbolic geometry, $K$ real), an imaginary conic (elliptic geometry, $K$ imaginary), or degenerate to two coincident lines and two imaginary points (parabolic or euclidean geometry, $K$ infinite).

**11.** In the last case there is a difficulty in determining the distance. Since $X$, $Y$ coincide, the cross-ratio $(PQ, XY)$ is zero and $K$ must be infinite; but the distance becomes now indeterminate.

Suppose $\qquad PY = PX + \epsilon$, where $\epsilon$ is small.

Then
$$(PQ, XY) = \frac{PX}{QX} \cdot \frac{QX + \epsilon}{PX + \epsilon} = \left(1 + \frac{\epsilon}{QX}\right)\left(1 + \frac{\epsilon}{PX}\right)^{-1}$$
$$= 1 + \epsilon\left(\frac{1}{QX} - \frac{1}{PX}\right),$$
neglecting squares and higher powers of $\epsilon$,

and
$$(PQ) = K \log (PQ, XY) = K\epsilon\left(\frac{1}{QX} - \frac{1}{PX}\right).$$

Let $K \to \infty$ and $\epsilon \to 0$ in such a way that $K\epsilon \to$ a finite limit $\lambda$.

Then
$$(PQ) = \lambda \frac{PQ}{PX \,.\, QX}.$$

Now, to fix $\lambda$ we must choose a point $E$ such that $(PE)=1$, the unit distance.

Then $(PQ)=\frac{PX\,.\,EX}{PE}\cdot\frac{PQ}{PX\,.\,QX}=\frac{XE}{PE}\div\frac{XQ}{PQ}=(XP,\ EQ)$.

If we measure distances from $P=0$ as origin,

$$(0Q)=(X0,\ EQ)=(0\infty\ ,\ Q1)=\frac{0Q}{01}\div\frac{\infty Q}{\infty 1},$$

which agrees with the expression in euclidean geometry, since $\frac{\infty Q}{\infty 1}=1$, and $01=1$.

This case differs in one marked respect from the case of elliptic geometry. In that system there is a natural unit of length, which may be taken as the length of the complete straight-line—the period, in fact, of linear measurement; just as in ordinary angular measurement there is a natural unit of angle, the complete revolution. In euclidean geometry, however, the unit of length has to be chosen conventionally, the natural unit having become infinite.

**12.** The other geometries, in which the measure of angle is either hyperbolic or parabolic, are of a somewhat bizarre nature.

For example, if the absolute degenerates to two imaginary lines $\omega$, $\omega'$, and two coincident points $\Omega$, the case is just the reciprocal of the euclidean case; linear measurement is elliptic, $K$ being imaginary, and angular measurement is parabolic, $k$ being infinite. In this geometry the straight line is of finite length $=\pi K i$. If the positive direction along any one line is defined, the positive directions along all other lines in a plane are determined, for this is determined by the sense of rotation about the point $\Omega$. The sides of a triangle are defined as the segments which subtend the opposite angles which do not contain $\Omega$, just as in euclidean geometry the angles of a triangle are defined as the angles which are subtended by the opposite segments which do not cut the line at infinity.

Thus the sides of the triangle $ABC$ in the figure (Fig. 88) are represented by the heavy lines. If the positive direction on each line is then defined as the direction corresponding to clockwise rotation about $\Omega$, then

$$a+b+c=\text{the length of the complete line,}$$

*i.e. the perimeter of a triangle is constant and* $=\pi Ki$.

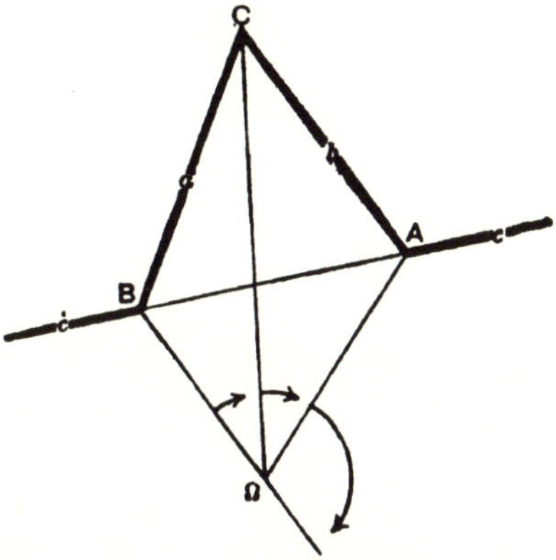


FIG. 88.

## 13. Extension to three dimensions.

In three dimensions the absolute is a quadric surface. If the measures of angle between lines and between planes are to be elliptic, the tangent planes through an actual line must be imaginary, and the tangents through an actual point in an actual plane must be imaginary.

(1) Let the quadric be real. If the quadric has real generators, *i.e.* if it is a ruled quadric, every plane cuts it in a real conic, for it cuts all the generators in real points. Actual points must lie within the section and actual lines must cut the surface. But the tangent planes passing through a line which cuts a ruled quadric are real, and so the measure of dihedral angles would be hyperbolic. The quadric cannot therefore be ruled.

Through a line which does not cut a non-ruled quadric

two real tangent planes pass; hence actual lines, and therefore planes, must cut the surface, and actual points are within the surface. This gives *hyperbolic geometry*. The absolute could be represented by a real sphere. All points outside the sphere are ideal points.

(2) Let the quadric be imaginary. The measure of distance is also elliptic, and the geometry is *elliptic*.

(3) Let the quadric degenerate. If the quadric degenerates to a cone, necessarily with real vertex, the measure of dihedral angle must be parabolic. If the quadric degenerates to two planes, unless the planes coincide they will have a real line of intersection and the measure of plane angle .must be parabolic. Hence the quadric must degenerate to two coincident planes.

A quadric which reduces, as a locus, to two coincident planes, reduces as an envelope to a conic lying in this plane. If the measure of angle is elliptic this conic must be imaginary. This is the case of *euclidean geometry*. The absolute consists of an imaginary conic in the plane at infinity. Any quadric which passes through this conic is cut by every plane in a conic which passes through the two absolute, or circular, points in this plane, *i.e.* every plane section is a circle, and the quadric is a sphere. The imaginary conic itself must be regarded as a circle since it is the plane section of a sphere. This is the *imaginary circle at infinity*.

**14.** Other three-dimensional geometries can be constructed in which the measure of plane or dihedral angle is hyperbolic or parabolic, but they are not of much interest, as they resemble ordinary geometry too slightly.

One application of these bizarre geometries may be given. It is obvious that in euclidean space the geometry on the plane at infinity is elliptic, since the absolute consists of the imaginary circle in this plane, and it follows, as we have already seen, that the geometry of complete straight lines through a point is elliptic, the geometry of rays, or of points on a sphere, being of course the spherical or antipodal variety.

Now consider three-dimensional hyperbolic space. A tangent plane to the absolute cuts the absolute in a degenerate conic consisting of two imaginary straight lines and two coincident points $\Omega$; hence the geometry on such a plane is the reciprocal of euclidean, *i.e.* the measure of distance is elliptic, while angular measurement is parabolic. Now, the polar of a point (or line) on this plane is a plane (or line) passing through $\Omega$. Hence, by this second reciprocation, we find that the geometry of a bundle of parallel lines and planes is euclidean, and if we cut the system by the surface (horosphere) which cuts each line and plane orthogonally, we find that *the geometry on the horosphere is euclidean.*

## Geodesic Representation.

**15.** It has been seen that the trigonometrical formulae of elliptic geometry with constant $k$ are exactly the same as those of spherical trigonometry on a sphere of radius $k$, and therefore elliptic geometry can be truly represented on a sphere, straight lines being represented by great circles, and antipodal points being regarded as identical. Within a limited region of the sphere which contains no pair of antipodal points, the geometry is exactly the same as elliptic geometry. We do not require, as in Cayley's representation, to obtain a distance- or angle-function; distances and angles are represented by the actual distances and angles on the sphere.

The corresponding representation for hyperbolic geometry appears at first sight to be imaginary, since hyperbolic geometry is the same as the geometry upon a sphere of purely imaginary radius. It is possible, however, to obtain a real representation of this kind, though confined to a limited portion of the hyperbolic plane.

## 16. Geometry upon a curved surface.

We must first understand what we mean by the geometry upon a surface which is not, like the sphere, uniform. The straight line joining two given points has the property that the distance measured along it is less than that measured along any other line joining the same two points. This is the property which we shall retain upon the surface. A curve lying on a surface and having this minimum property is called a *geodesic*. The geodesics of a sphere are all great circles. Now, if a surface can be bent in any way, without stretching, creasing or tearing, geodesics will remain geodesics, lengths of lines and magnitudes of angles will remain unaltered, and the geometry on the surface remains precisely the same. Two surfaces which can be transformed into one another in this way are called *applicable* surfaces.

If, for example, a plane is bent into the form of a cylinder, the geometry, at least of a limited region, will be unaltered. The same holds for any surface which can be laid flat or *developed* on the plane.

The sphere is a surface which cannot be developed on the plane, and it possesses a geometry of its own. A complete sphere cannot in fact be bent at all without either stretching or kinking, but a limited portion of it can be bent into different shapes without altering the character of the geometry.

## 17. Measure of curvature.

Now the sphere and the plane possess this property in common, that congruent figures, *e.g.* triangles with equal corresponding sides and angles, can be constructed in any positions on the surface, or, to use the language of kinematics, a rigid figure is freely movable on the surface. It follows that the surface is applicable to itself at all its points. This property is expressed analytically by saying that there is a certain quantity, called the measure of curvature, which is the same at all points of the surface and is not altered by bending.

To see what this invariant quantity is, consider any plane section of a surface passing through a tangent line $OT$ at $O$; the section is a curve having this line as tangent at $O$. The more obliquely the plane cuts the surface the sharper is the curvature of the section, until, when the plane touches the surface at $O$, the section is just a point.[1] The section of least curvature occurs when the plane is perpendicular to the tangent plane, or passes through the normal to the surface.

Again, if we revolve the cutting plane about the normal, the curvature of the section will vary continuously and have a maximum and a minimum value. These occur for sections at right angles, and are called the *principal curvatures* of the surface at $O$. The curvature of a curve at a point $O$ being defined as the reciprocal of the radius of the circle of closest fit to the curve at $O$, the product of the principal curvatures at $O$, denoted by $M$, is called the *measure of curvature* of the surface at $O$. If the two curvatures are in the same sense $M$ is positive, if in opposite senses $M$ is negative; if one is zero, as in the case of a cylinder or any developable surface, $M$ is zero. For a sphere of radius $k$, $M$ is the same at all points and $=1/k^2$.

[1] This holds for a convex surface like a sphere. In the general case the section of a surface by a tangent plane is a curve which has a node at the point of contact, with real or imaginary tangents. In the case of a surface of the second degree the section consists of two straight lines, real or imaginary. In the text we are considering the case of a node with imaginary tangents, which appears just as a point.

### 18. Surfaces of constant curvature.

Gauss, who founded the differential geometry of surfaces, as well as being almost the discoverer of non-euclidean geometry, discovered the beautiful theorem [1] that *when a surface is bent in any way without stretching or kinking, the measure of curvature at every point remains unaltered.* It follows, then, that the only surfaces upon which free mobility is possible are those which are applicable upon themselves in all positions, and therefore for which $M$ has the same value at all points.

There are three kinds of surfaces of constant curvature, (1) those of constant positive curvature, of which the sphere is a type; (2) those of constant negative curvature, saddle-backed at all points like a " diabolo "; (3) those of zero curvature, the plane and all developables.

### 19. The pseudosphere.

Fortunately, we do not require to take an imaginary sphere as the type of surfaces of constant negative curvature. There are different forms of such surfaces, even of revolution, but the simplest is the surface called the *Pseudosphere*, which is formed by revolving a tractrix about its asymptote.

The tractrix is connected with the simpler curve, the catenary, which is the form in which a uniform chain hangs under gravity. The equation of the catenary referred to the axes $Ox$, $Oy$ is $y = k \cosh \frac{x}{k}$. It has the property that the distance of the foot of the ordinate $N$ from the tangent at $Q$ is constant and equal to $k$, while $QP =$ the

[1] C. F. Gauss, *Disquisitiones generales circa superficies curvas*, Göttingen, 1828 (§ 12).

arc $AQ$. It follows then that if a string $AQ$ is unwound from the curve, its extremity will describe a curve $AP$ with the property that the length of the tangent $PN$ is constant and equal to $k$. This curve is the Tractrix. $Ox$ is an asymptote. Now, if the tractrix is revolved about the asymptote we get a surface of revolution whose principal

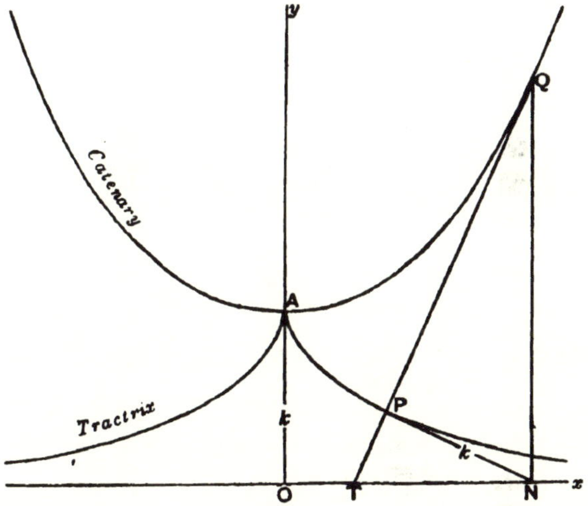


FIG. 89.

sections at $P$ are the meridian section in which the tractrix lies, and a section through the normal $PT$ at right angles to the plane of the curve. The radii of curvature of these sections are respectively $PQ$ and $PT$, and we have $PQ \, . \, PT = PN^2 = k^2$, but since the curvatures are in opposite senses, the measure of curvature $= -1/k^2$.

The pseudosphere, therefore, gives a real surface upon which hyperbolic geometry is verified—within a limited region. The surface does not, of course, represent the whole of the hyperbolic plane, for it has only a single point at infinity. The meridian curves are geodesics passing through this point at infinity, and therefore represent a system of parallel lines. So the surface only corresponds

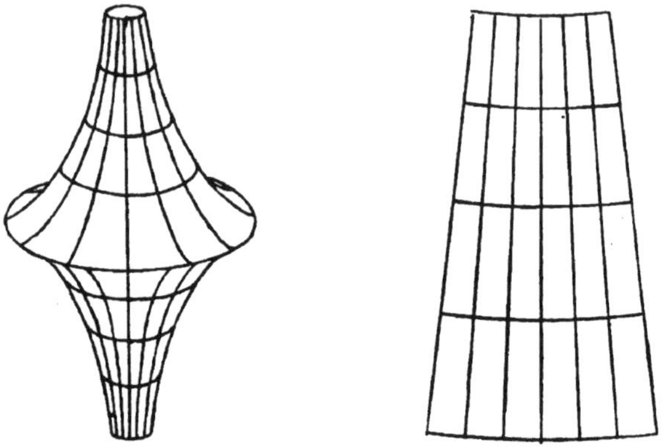

FIG. 90.

to a portion of the hyperbolic plane bounded by two parallel lines and an arc of a horocycle.[1]

## 20. The Cayley-Klein representation as a projection.

Through the medium of the geodesic representation we can now get a geometrical interpretation of the Cayley-Klein representation. If we project a sphere centrally, great circles are projected into straight lines, since their planes pass through the centre of projection. Hence the Cayley-Klein representation of elliptic geometry can be regarded as the central or gnomonic projection of the geometry on a sphere.

The equations of transformation are easily found.

[1] The intrinsic equation of the tractrix is $s=k\log\operatorname{cosec}\psi$, and since $y=k\sin\psi$ we have $y=ke^{-s/k}$. The ratio of the corresponding arcs of two horocycles (sections $\perp$ to the axis) is therefore $e^{(s-s')/k}$, which agrees with the expression we have already found (Chap. II. § 26).

Let the plane of projection be chosen for convenience as the tangent plane at $O$, and take rectangular axes, $Oz$ through the centre of the sphere $S$, and $Ox$, $Oy$ in the plane

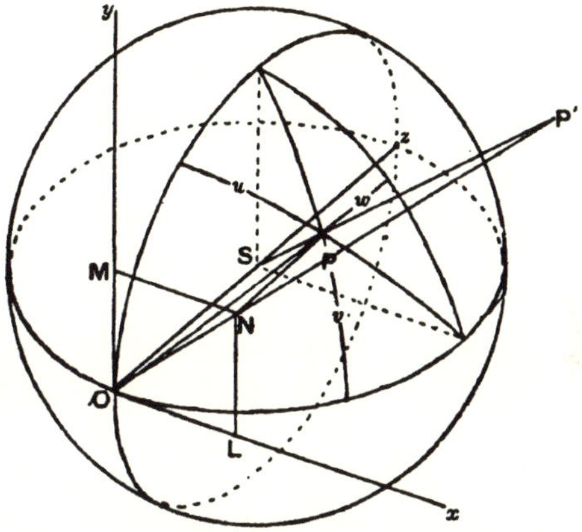


FIG. 91.

of projection. Let the coordinates of $P$, any point on the sphere, be $(x, y, z)$ and the coordinates of its projection $P'$ be $(x', y', 0)$. Then $x^2+y^2+(z-k)^2=k^2$,

$$\frac{x'}{x}=\frac{y'}{y}=\frac{OP'}{ON}=\frac{OS}{OS-NP}=\frac{k}{k-z}$$

$$=\frac{k}{\sqrt{k^2-x^2-y^2}}=\frac{\sqrt{x'^2+y'^2+k^2}}{k}.$$

## 21. Meaning of Weierstrass' coordinates.

Let $u$, $v$, $w$ be the angles which $SP$ makes with the planes $yz$, $zx$ and $xy$; $u$, $v$, $w$ can be regarded as coordinates on the sphere, and

$$x=k\sin u, \quad y=k\sin v, \quad z-k=k\sin w.$$

Then $k\sin u$, $k\sin v$ and $\sin w$ are Weierstrass' coordinates, denoted by $X$, $Y$, $Z$, and connected by the relation $X^2+Y^2+k^2Z^2=k^2$. In terms of the spatial coordinates of $P$, $X=x$, $Y=y$, $kZ=z-k$.

We see thus that Weierstrass' coordinates are proportional to the sines of the distances of $P$ from the sides of a self-polar triangle, and are therefore analogous to trilinear coordinates.

Dually, the line-coordinates of a line are defined as proportional to the sines of the distances of the line from the vertices of the fundamental triangle. We may also define the point-coordinates as proportional to the cosines of the distances from the vertices, and the line-coordinates as proportional to the cosines of the angles which the line makes with the sides of the triangle. In the three-dimensional representation the point-coordinates are the direction-sines of the point referred to rectangular axes.

## Conformal Representation.

### 22. Stereographic projection.

There is another very useful projection of a sphere, the stereographic projection. In this case the centre of projection is taken on the surface.

Let $S$ be the centre of projection, and $C$ the centre of the sphere of radius $k$. Take the plane of projection perpendicular to $SC$, and at distance $SO=d$. Choose rectangular axes with $OS$ as axis of $z$. Let the coordinates of $P$, any point on the sphere, be $(x, y, z)$, and the coordinates of its projection $P'$ be $(x', y', 0)$. Then, since $SPA$ is a right angle $=SOP'$, $SP \,.\, SP'=SA \,.\, SO=2kd$.

The formulae of transformation are:

$$\frac{x'}{x}=\frac{y'}{y}=\frac{OP'}{ON}=\frac{SP'}{SP}=\frac{d}{d-z}=\frac{2kd}{x^2+y^2+(z-d)^2}=\frac{x'^2+y'^2+d^2}{2kd}.$$

If the plane of projection is chosen to pass through $C$, and the $xy$ plane of $P$ is the tangent plane at $S$, $d=k$, and the formulae become:

$$\frac{x'}{x}=\frac{y'}{y}=\frac{k}{z}=\frac{2k^2}{x^2+y^2+z^2}=\frac{x'^2+y'^2+k^2}{2k^2}.$$

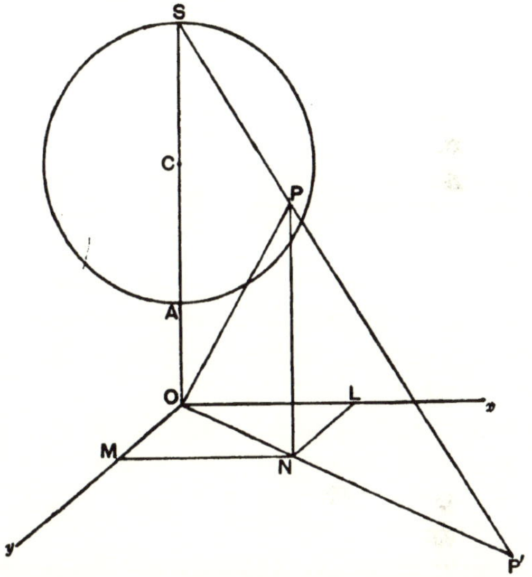


FIG. 92.

A plane $ax+by+cz+d=0$ becomes

$$2k^2(ax'+by'+ck)+d(x'^2+y'^2+k^2)=0,$$

which represents a circle. Hence *all circles on the sphere are represented by circles.*

Consider two planes

$$lx+my+nz=nk, \quad l'x+m'y+n'z=n'k$$

through the centre, and cutting the sphere in great

circles. The angle between the great circles is equal to the angle between the planes, and is given by

$$\cos\theta = ll' + mm' + nn'.$$

The projections of the great circles are the circles

$$2k^2(lx + my + nk) - nk(x^2 + y^2 + k^2) = 0,$$

*i.e.*
$$\left(x - \frac{kl}{n}\right)^2 + \left(y - \frac{km}{n}\right)^2 = \frac{k^2}{n^2}, \text{ etc.}$$

The angle $\phi$ at which the circles cut is given by

$$k^2\left(\frac{l}{n} - \frac{l'}{n'}\right)^2 + k^2\left(\frac{m}{n} - \frac{m'}{n'}\right)^2 = \frac{k^2}{n^2} + \frac{k^2}{n'^2} - 2\frac{k^2}{nn'}\cos\phi;$$

therefore
$$\cos\phi = ll' + mm' + nn',$$

*i.e. the projections cut at the same angle as the great circles.*

Stereographic projection is three-dimensional inversion, for $SP \,.\, SP' = \text{const.}$, circles are changed into circles, and angles are unaltered. A small figure on the sphere will therefore be projected into a similarly shaped small figure on the plane, with corresponding angles all equal. For this reason the representation is called *conformal.*

## 23. The orthogonal circle or absolute.

A circle in the projection, which represents a straight line in the non-euclidean geometry, has for its equation

$$n(x^2 + y^2) - 2k(lx + my) - nk^2 = 0,$$

and this cuts at right angles the fixed circle

$$x^2 + y^2 + k^2 = 0.$$

This circle, imaginary in elliptic geometry, real in hyperbolic, is the projection of the absolute, which cuts all straight lines at right angles.

## 24. Conformal representation.

We shall now consider generally the problem of the conformal representation of the non-euclidean plane upon the euclidean plane, straight lines being represented by circles. We shall set aside stereographic projection entirely, as this assumes the geodesic representation on a sphere, and treat the problem purely as a problem in correspondence.

It may be shown directly that the circles which represent straight lines all cut a fixed circle orthogonally. For, if the circles

$$x^2+y^2+2gx+2fy+c=0$$

represent straight lines, they must have the property of being determined uniquely by two points. Hence the three coefficients $g$, $f$, $c$ must be connected by a fixed equation of the first degree, say

$$2gg'+2ff'=c+c',$$

but this is just the condition that the circle should cut orthogonally the fixed circle

$$x^2+y^2+2g'x+2f'y+c'=0.$$

In elliptic geometry this circle is imaginary, in hyperbolic geometry it is real. If the fixed circle reduces to a point (the transition between a real and an imaginary circle), all the circles which represent straight lines pass through this fixed point. Now, if we invert the system with this point as centre, the circles become straight lines. Hence the straight lines of euclidean geometry can be represented by a system of circles passing through a fixed point.

## 25. Point-pairs.

This representation has a fault which we must try to correct. In non-euclidean geometry, hyperbolic or elliptic, two straight lines intersect in only one point; but the circles which represent them intersect in a pair of points.

In the representation of hyperbolic geometry the fixed circle is real, and two orthogonal circles may intersect in real, imaginary or coincident points, according as the straight lines which they represent are intersectors, non-intersectors or parallel. The pair of points which correspond to a single point are inverses with respect to the fixed circle. We must therefore consider pairs of points which are inverses with respect to the fixed circle as forming just one point.

In the representation of spherical geometry, as distinct from elliptic geometry, the points of a pair will be considered as distinct and constituting a pair of antipodal points.

In the representation of euclidean geometry, one of the points of a pair is always the fixed point itself.

### 26. Pencils of lines. Concentric circles.

To a pencil of lines through a point $P$ corresponds a pencil of circles through the two points $P_1$, $P_2$ which correspond to $P$. The radical axis of this system is itself a circle of the system, and is in no way distinguished from any other circle of the system.

The representation of straight lines by circles is not necessarily a conformal one, nor is it by any means the only possible conformal representation. If the representation is conformal we can show that when straight lines are represented by circles, *circles also are represented by circles.*[1] For a system of concentric circles cut all the lines of a pencil with vertex $P$ at right angles. They will therefore be

[1] For the converse of this theorem, see Chap. VIII. § 2.

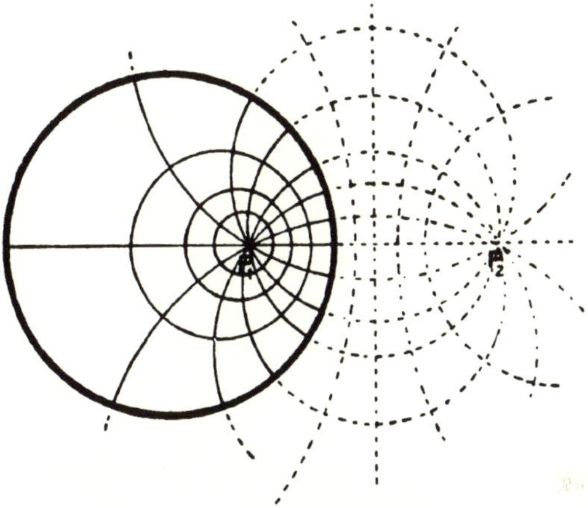

FIG. 93.

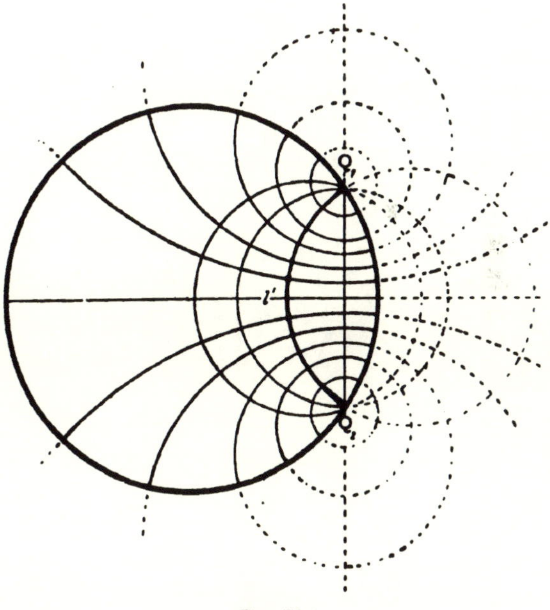

FIG. 94.

represented by a system of curves cutting orthogonally a system of coaxal circles; but this is also a system of coaxal circles, and has $P_1$, $P_2$ as limiting points (Fig. 93).

A circle is represented actually by a pair of circles; these are inverses with respect to the fixed circle, and are coaxal with the fixed circle. The two limiting points form its centre.

Corresponding to a pencil of lines through an ideal point $P$, *i.e.* a system of lines cutting a fixed line $l$ at right angles, we have a system of circles cutting at right angles the fixed circle and the circle $l'$ which represents the fixed line (Fig. 94). But this is a coaxal system whose radical axis is the common chord of the fixed circle and the circle $l'$; its limiting points are the common points $Q_1$ and $Q_2$ of $l'$ and the fixed circle. The circles with centre $P$ are equi-distant-curves, and are represented by a system of circles passing through $Q_1$, $Q_2$.

Corresponding to a pencil of lines through a point at

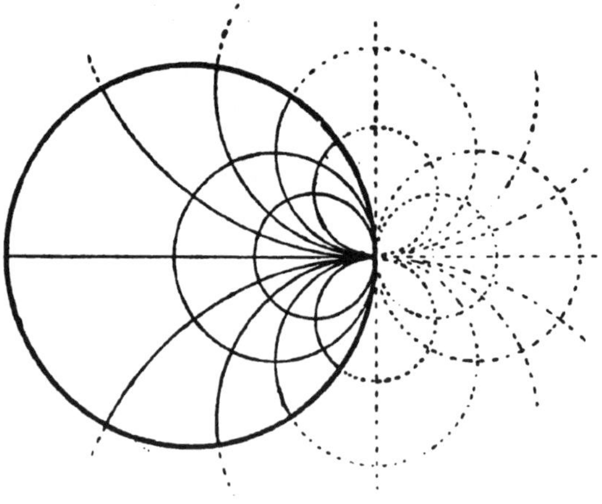

FIG. 95.

infinity $P$, *i.e.* a system of parallel lines, we have a system of circles cutting the fixed circle orthogonally at a fixed point on it (Fig. 95). The horocycles, circles with centre $P$, are represented by circles touching the fixed circle at this point.

**27. The distance between two points.**

If $ABC$ is a circle which represents a circle in non-euclidean geometry, and $O$ is the point which represents

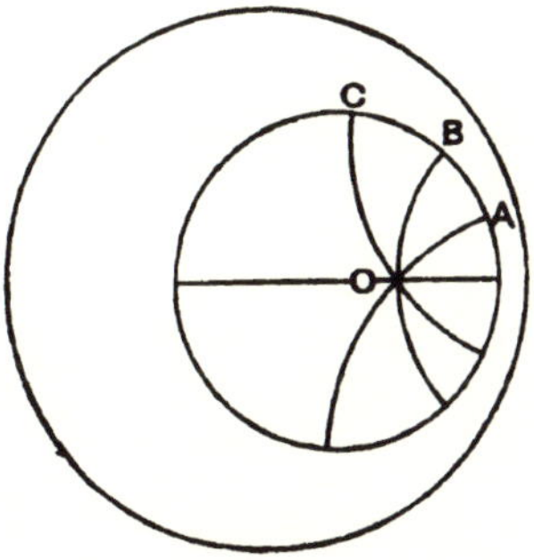


FIG. 96.

its centre, the radii are represented by arcs of circles through $O$ cutting the given circle and the fixed circle orthogonally. The arcs $OA$, $OB$, $OC$, ... represent equal distances in non-euclidean geometry. We require then to find what function of the positions of the points $O$ and $A$ represents the distance between their corresponding points.

**28. Motions.**

In order to investigate this function we shall make use of the idea of motion. By a motion or displacement in the general sense is meant not a change in position of a single point or of any bounded figure, but a displacement of the whole space, or, if we are dealing only with two dimensions, of the whole plane. A motion is a transformation which changes each point $P$ uniquely into another point $P'$ in such a way that distances and angles are unchanged. It follows that straight lines remain straight lines, and the displacement is a particular case of a *collineation* (the general one-one point-transformation which changes straight lines into straight lines). Further, it will change circles into circles, and the fixed circle must remain fixed as a whole. We require therefore to find what is the sort of transformation of the euclidean plane which will change circles into circles and leave a fixed circle unaltered.

## 29. Reflexions.

The process of inversion with respect to a circle at once suggests itself, since this transformation leaves angles unaltered and changes circles into circles. Further, since

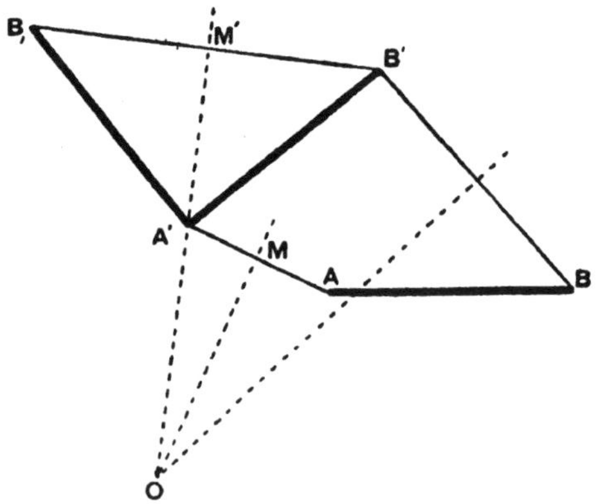


FIG. 97.

the fundamental circle must be unaltered as a whole, the circle of inversion must cut it orthogonally. Let us then consider inversion in a circle which represents a straight line.

In euclidean geometry, when the circle of inversion becomes a straight line, inversion reduces to *reflexion* in this line. Now *any motion or displacement in euclidean geometry can be reduced to a pair of reflexions in two suitably chosen lines.*

If $AB$ is displaced to $A'B'$ (Fig. 97), first take $MO$ the perpendicular bisector of $AA'$; the reflexion of $AB$ in $MO$ is $A'B_1$. Then take $M'O$ the perpendicular bisector of $B_1B'$, which passes through $A'$, and the reflexion of $A'B_1$ is $A'B'$. Since $OB = OB_1 = OB'$, $O$ lies also on the perpendicular bisector of $BB'$, and is in fact the centre of rotation for the given displacement.

## 30. Complex numbers.

We shall find now what is the most general transformation which changes circles into circles and the fundamental circle into itself.

The equation of any circle is

$$x^2 + y^2 + 2gx + 2fy + c = 0.$$

The procedure is greatly simplified by the introduction of complex numbers and the use of Argand's diagram. Let $z = x + iy$, $p = g + if$, and write the conjugate complex numbers $\bar{z} = x - iy$, $\bar{p} = g - if$. Then the equation of the circle becomes

$$z\bar{z} + \bar{p}z + p\bar{z} + c = 0, \quad \ldots\ldots\ldots\ldots\ldots (1)$$

a lineo-linear expression in $z$, $\bar{z}$. Its centre is $z = -p$, and the square of its radius is $p\bar{p} - c$.

**31. Circular transformation, conformal and homographic.**

Now it is proved in the theory of functions that any transformation of the form

$$z=f(z'),\quad \bar{z}=f(\bar{z}')$$

is conformal, leaving angles unchanged. A real transformation of this form which leaves the form of the equation (1) unaltered, *i.e.* which changes circles into circles, is one in which $z$, $z'$ both occur only to the first power,[1] or

$$z=\frac{\alpha z'+\beta}{\gamma z'+\delta},\quad \bar{z}=\frac{\bar{\alpha}\bar{z}'+\bar{\beta}}{\bar{\gamma}\bar{z}'+\bar{\delta}},$$

where $\alpha$, $\beta$, $\gamma$, $\delta$ are any complex numbers such that $\alpha\delta \neq \beta\gamma$.

By this transformation any complex number $z$ is transformed into a complex number $z'$, and the point $(x, y)$ corresponding to $z$ is transformed into the point $(x', y')$ corresponding to $z'$.

If $z_1, z_2, z_3, z_4$ are any four complex numbers which are transformed into $z_1', z_2', z_3', z_4'$, and if we define the cross-ratio $(z_1z_2, z_3z_4)$

$$=\frac{z_1-z_3}{z_1-z_4}\div\frac{z_2-z_3}{z_2-z_4},$$

then

$$(z_1z_2, z_3z_4)=(z_1'z_2', z_3'z_4').$$

The transformation is therefore said to be *homographic.*

Let $r_{13}$ be the modulus, and $\theta_{13}$ the amplitude of the complex number $z_1-z_3$, so that $z_1-z_3=r_{13}e^{i\theta_{13}}$, then the cross-ratio $(z_1z_2, z_3z_4)$ has modulus $\dfrac{r_{13}}{r_{14}}\Big/\dfrac{r_{23}}{r_{24}}$ and amplitude $\theta_{13}-\theta_{14}-\theta_{23}+\theta_{24}$ or $\theta_{13}+\theta_{32}+\theta_{24}+\theta_{41}$.

[1] The only other type of real transformation having this property is that which differs only from this one by interchanging $z$ and $\bar{z}$. But this only differs from the former by a reflexion in the $x$-axis, $z=\bar{z}''$, $\bar{z}=z''$.

The cross-ratio is real only when its amplitude is a multiple of $\pi$, *i.e.* when the points corresponding to the four numbers $z_1$, $z_2$, $z_3$, $z_4$ are concyclic, and its value is then $\dfrac{P_1P_3}{P_1P_4} \div \dfrac{P_2P_3}{P_2P_4}$ (Fig. 98).

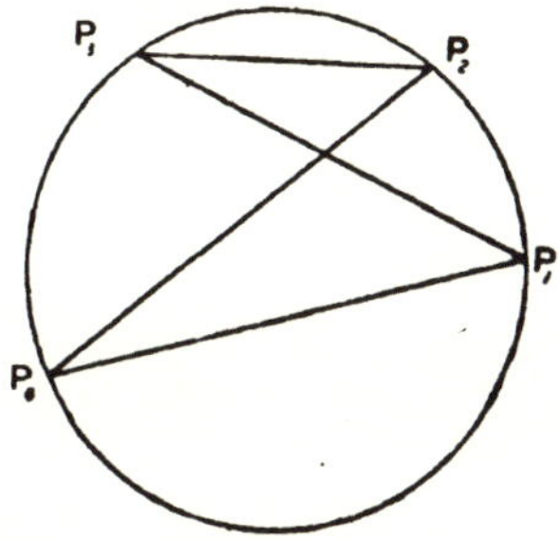


FIG. 98.

The transformation has to satisfy the further condition that it transforms the fundamental circle into itself.

It can be proved that if the fundamental circle is $x^2+y^2+K=0$, or $z\bar{z}+K=0$, the general form of the transformation is

$$z=\frac{az'-K\bar{\beta}}{\beta z'+a}.$$

If the fundamental circle is $y=0$, or $z=\bar{z}$, it can be proved that the general transformation is

$$z=\frac{az'+b}{cz'+d},$$

where $a$, $b$, $c$, $d$ are real and $ad \neq bc$.

## 32. Inversion.

Consider now the equations of transformation of inversion in a circle cutting the fundamental circle orthogonally.

The equation of a circle cutting $z\bar{z}+K=0$ orthogonally is $z\bar{z}+\bar{p}z+p\bar{z}-K=0$.

Let $C(-p)$ be the centre, $P(z)$ and $P'(z')$ a pair of inverse points.

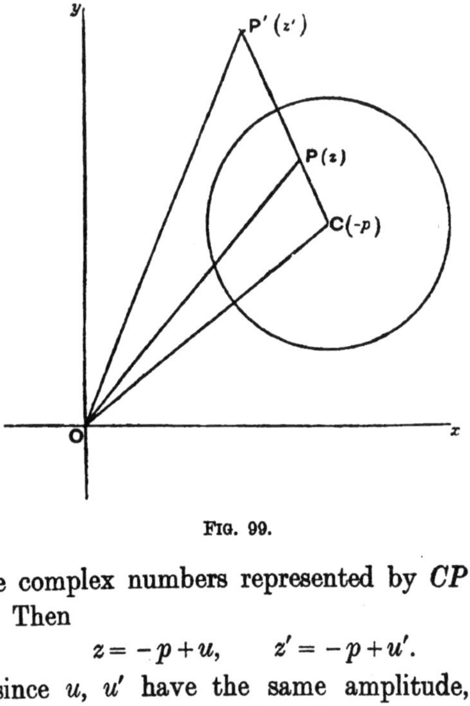


FIG. 99.

Let the complex numbers represented by $CP$ and $CP'$ be $u$, $u'$. Then

$$z=-p+u, \qquad z'=-p+u'.$$

Also, since $u$, $u'$ have the same amplitude, and the product of their moduli is equal to the square of the radius of the circle of inversion,

$$u\bar{u}'=p\bar{p}+K.$$

Therefore $$(z+p)(\bar{z}'+\bar{p})-p\bar{p}-K=0,$$

or $$z\bar{z}'+\bar{p}z+p\bar{z}'-K=0,$$

*i.e.* $$z=\frac{-p\bar{z}'+K}{\bar{z}'+\bar{p}}.$$

A second inversion in the circle $z\bar{z}+\bar{q}z+q\bar{z}-K=0$ gives

$$z=\frac{(K+p\bar{q})z''-K(p-q)}{(\bar{p}-\bar{q})z''+(K+\bar{p}q)}.$$

This will not hold when the circle of inversion is a straight line $\theta=\phi$. Here inversion becomes reflexion, and we have

$$z=re^{i\theta},\quad z'=re^{i(2\phi-\theta)},\quad \bar{z}'=re^{i(\theta-2\phi)};$$

therefore $$z=\bar{z}'e^{2i\phi}.$$

This combined with an inversion gives

$$z=\frac{-\bar{p}z''+K}{z''+p}e^{2i\phi}.$$

Let $\phi=\frac{\pi}{2}-\psi$, $\beta=e^{i\psi}$; then $e^{2i\phi}=-e^{-2i\psi}=-\frac{\bar{\beta}}{\beta}$. Then, if $p\beta=\alpha$, the transformation becomes

$$z=\frac{\bar{\alpha}z''-K\bar{\beta}}{\beta z''+\alpha}.$$

Hence in either case the transformation is of this form. Hence *the general displacement of a plane figure is equivalent to a pair of inversions in two circles which cut the fundamental circle orthogonally.*

## 33. Types of motions.

In the general displacement there are always two points which are unaltered, for if $z'=z$ we have the quadratic equation

$$\beta z^2+(\alpha-\bar{\alpha})z+K\bar{\beta}=0.$$

If we substitute $z=-K/\bar{z}'$, the equation becomes

$$\bar{\beta}\bar{z}'^2+(\bar{\alpha}-\alpha)\bar{z}'+K\beta=0;$$

therefore $z'$ is also a root. The two points are therefore inverses with regard to the fundamental circle. This point-pair corresponds to the centre of rotation in the general displacement. In hyperbolic geometry there are three distinct types of displacement according as the centre of rotation is real, ideal, or at infinity. The first case is similar to ordinary rotation; the second case is motion of translation along a fixed line, and points not on this line describe equidistant-curves; in the third case all points describe arcs of horocycles.

## 34. The distance-function.

We have now to find the expression for the distance between two points $P$, $Q$, *i.e.* the function of their coordinates or complex numbers $(z_1, z_2)$, which remains invariant during a motion.

The two points determine uniquely a circle cutting the fundamental circle orthogonally in $X$, $Y$. This circle represents the straight line joining $PQ$, and $X$, $Y$ represent the points at infinity on this line. If the motion is one of translation along this line, the straight line as well as the fundamental circle are unaltered, and $X$, $Y$ are fixed points. Let $x$, $y$ be the complex numbers corresponding to $X$, $Y$; then the cross-ratio $(z_1z_2, xy)$ remains constant. If we suppose, therefore, that for points on this line the distance $(PQ)$ is a function of this cross-ratio, we can write $(PQ)=f(z_1, z_2)$. If $P$, $Q$, $R$ are three points on the line, corresponding to the numbers $z_1$, $z_2$, $z_3$, this function has to satisfy the relation $(PQ)+(QR)=(PR)$, or

$$f(z_1, z_2)+f(z_2, z_3)=f(z_1, z_3).$$

This is a functional equation by which the form of the function is determined. Consider $z$ as a parameter determining the position of a point, and differentiate with regard to $z_1$. Then, since

$$(z_1z_2, xy)=\frac{z_1-x}{z_1-y}\cdot\frac{z_2-y}{z_2-x}=\frac{PX}{PY}\cdot\frac{QY}{QX},$$

we have

$$f'(z_1, z_2)\frac{QY}{QX}\frac{\partial}{\partial z_1}\left(\frac{PX}{PY}\right)=f'(z_1, z_3)\frac{RY}{RX}\frac{\partial}{\partial z_1}\left(\frac{PX}{PY}\right).$$

Hence

$$\frac{f'(z_1, z_2)}{f'(z_1, z_3)}=\frac{QX}{QY}\cdot\frac{RY}{RX}=\left(\frac{PX}{PY}\cdot\frac{RY}{RX}\right)\div\left(\frac{PX}{PY}\cdot\frac{QY}{QX}\right)=\frac{(z_1z_3,\ xy)}{(z_1z_2,\ xy)},$$

*i.e.* $(z_1, z_2)f'(z_1, z_2)=(z_1, z_3)f'(z_1, z_3)=\text{const.}=\mu.$

Integrating, we find

$$f(z_1, z_2)=\mu\log(z_1z_2,\ xy)+C,$$

and substituting in the functional equation we find $C=0$.

Hence

$$(PQ)=\mu\log(z_1z_2,\ xy)=\mu\log\left(\frac{PX}{PY}\cdot\frac{QY}{QX}\right)=\mu\log(PQ,\ XY),$$

$(PQ, XY)$ being the cross-ratio of the four points $P$, $Q$, $X$, $Y$ on the circle, *i.e.* the cross-ratio of the pencil $O(PQ, XY)$, where $O$ is any point on the circle.

In hyperbolic geometry $K=-k^2$, and the fundamental circle is real. The distance between two conjugate points is $\frac{1}{2}i\pi k$, and the cross-ratio $(PQ, XY)=-1$. Then

$$(PQ)=\mu i\frac{\pi}{2}.$$

Therefore $\mu=k$.

## 35. The line-element.

If, returning to the stereographic projection, we take the formulae in § 22, we can find an expression for the line-element $ds$. We have, $x$, $y$, $z$ being the coordinates of a point on the sphere,

$$ds^2=dx^2+dy^2+dz^2.$$

Expressing this in terms of $x'$ and $y'$, we get

$$ds^2=\frac{4k^2d^2(dx'^2+dy'^2)}{(x'^2+y'^2+d^2)^2}.$$

In particular, if $d=2k$, so that the plane of projection is the tangent plane at $A$ (Fig. 92), we get

$$ds=\sqrt{dx'^2+dy'^2}/\{1+\tfrac{1}{4}a(x'^2+y'^2)\},$$

where $a=1/k^2$.

**36.** There is a gain in simplicity when the fundamental circle is taken as a straight line, say the axis of $x$. Then straight lines are represented by circles with their centres on the axis of $x$. Pairs of points equidistant from the axis

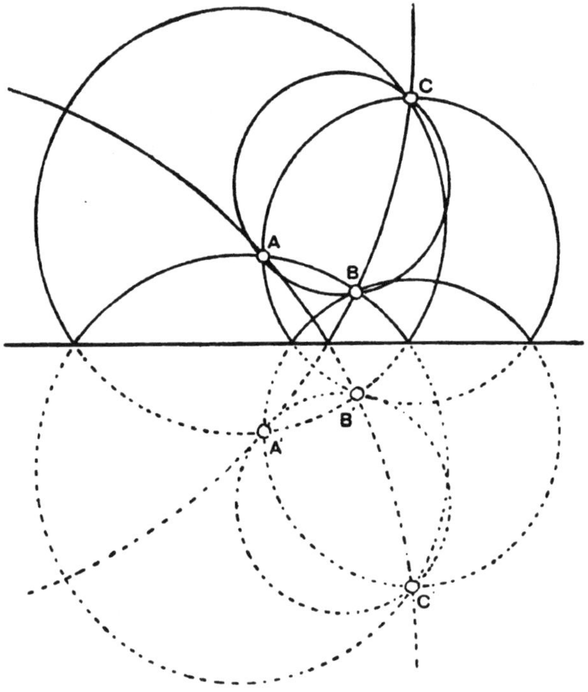


FIG. 100.

of $x$ represent the same point, and we may avoid dealing with pairs of points by considering only those points above the $x$-axis. A proper circle is represented by a circle lying entirely above the $x$-axis ; a horocycle by a circle touching the $x$-axis ; an equidistant-curve by the upper part of a circle cutting the $x$-axis together with the reflexion of the part which lies below the axis.

Through three points $A$, $B$, $C$ pass four circles. If $A'$, $B'$, $C'$ are the reflexions of $A$, $B$, $C$, the four circles are represented by $ABC$, $A'BC$, $AB'C$, $ABC'$. The last three are certainly equidistant-curves ; the first may be a proper circle, a horocycle or an equidistant-curve.

## 37. Angle at which an equidistant-curve meets its axis.

Fig. 100 shows that the two branches of an equidistant-curve cut at infinity at a finite angle, a fact that is not apparent in the Cayley-

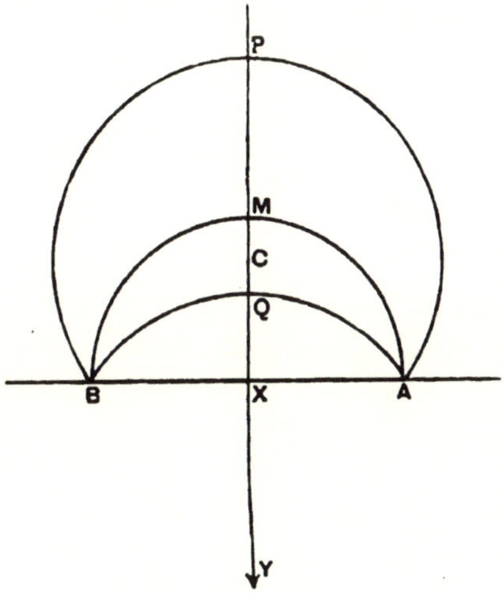


FIG. 101.

Klein representation. Let $APBQA$ (Fig. 101) be the equidistant-curve, $AMB$ its axis, represented by the circle on $AB$ as diameter, and let $C$ be the centre of the circle $APB$. Draw $CX \perp AB$ meeting the two branches of the equidistant-curve and its axis in $P, Q, M$.

Let $PAQ=2\alpha$; then $CAX=\alpha$, $\tan\alpha=\frac{CX}{XA}$. Let $d$ be the distance of the equidistant-curve from its axis.

The line $PX$ being $\perp AB$ represents a straight line; it cuts $AB$ in $X$, and the second point at infinity on the line is represented by $Y$ at infinity.

Hence $d=k\log(PM,\ XY)=k\log\frac{PX}{MX}$, also $d=k\log\frac{MX}{QX}$.

Now $$CX=\tfrac{1}{2}(PX-QX);$$

therefore $$\tan\alpha=\frac{PX-QX}{2MX}=\frac{1}{2}\left(e^{\frac{d}{k}}-e^{-\frac{d}{k}}\right)=\sinh\frac{d}{k}.$$

We can get a geometrical meaning for this result. Draw $PL \perp PN$ and $PE \parallel NE$ (Fig. 102). Then the equidistant-curve and the parallel and the axis all meet at infinity at $E$.

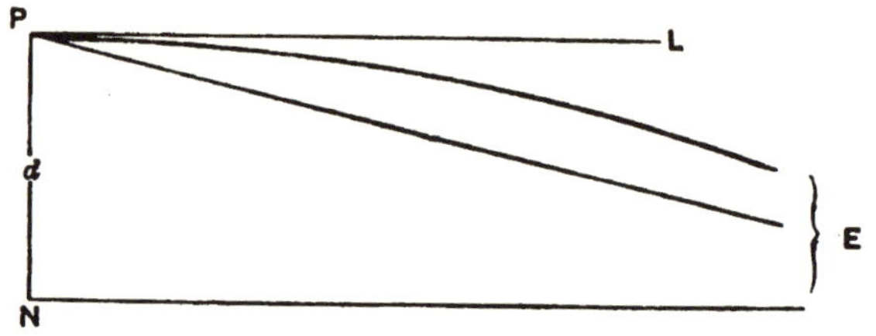


FIG. 102.

The angle $LPE$ is then $=\alpha$.

Consider a chord $PQ$ of the equidistant-curve: like a circle, the curve cuts the chord at equal angles. Keeping $P$ fixed, let $Q$ go to infinity. $PQ$ becomes parallel to $NE$, and makes a zero angle with it; hence the angle between the curve and the axis is equal to the angle $LPE$.

The explanation of the apparent contradiction shown in the Cayley-Klein representation, where the two branches of the equidistant-curve form one continuous curve, lies in the fact that the angle between two lines becomes indeterminate when their point of intersection is on the absolute and at the same time one of the lines touches the absolute. If the first alone happens the angle is zero, if the second the angle is infinite.

**38. Extension to three dimensions.**

The conformal representation of non-euclidean geometry can be extended to three dimensions, planes being represented by spheres cutting a fundamental sphere orthogonally. A proper sphere is represented by a sphere which does not cut the fundamental sphere, a horosphere by a sphere touching the fundamental sphere, and an equidistant-surface by a sphere cutting the fundamental sphere.

A horocycle is represented by a circle touching the fundamental sphere. The horocycles which lie on a horosphere all pass through the same point on the sphere, viz. the point of contact. This is exactly similar to the system of circles on a plane representing the straight lines of euclidean geometry, and thus we have another verification that the geometry on the horosphere is euclidean.

This suggests that the three geometries can be represented on the plane of any one of them by systems of circles cutting a fixed circle orthogonally.

# CHAPTER VI.

## "SPACE CURVATURE" AND THE PHILOSOPHICAL BEARING OF NON-EUCLIDEAN GEOMETRY.

### 1. Four periods in the history of non-euclidean geometry.

The projective and the geodesic representations of non-euclidean geometry have an important bearing on the history of the subject, for it was through these that Cayley and Riemann arrived independently at non-euclidean geometry.

Klein has divided the history of non-euclidean geometry into three periods. The *first period*, which contains GAUSS, LOBACHEVSKY and BOLYAI, is characterised by the synthetic method, and applies the methods of elementary geometry. The *second period* is related to the geodesic representation, and employs the methods of differential geometry. It begins with RIEMANN's classical dissertation, and includes also the work of HELMHOLTZ, LIE and BELTRAMI on the formula for the line-element. The *third period* is related to the projective representation, and applies the principles of pure projective geometry. It begins with CAYLEY, whose ideas were developed and put into relationship with non-euclidean geometry by KLEIN. To these a *fourth period* has now to be added, which is connected with no representation, but is concerned with the logical grounding of geometry upon sets of axioms. It is inaugurated by PASCH, though we must go back to VON STAUDT for the true beginnings. This period contains HILBERT and an Italian school represented by PEANO and PIERI; in America its chief representative is VEBLEN. It has led to the severe logical examination of the foundations of mathematics represented by the *Principia Mathematica* of RUSSELL and WHITEHEAD.

## 2. "Curved space."

If we attempt to extend the geodesic representation of non-euclidean geometry to space of three dimensions, we find ourselves at a loss, for the representation of plane geometry already requires three dimensions. It is quite a legitimate mathematical conception, however, to extend space to four dimensions. A limited portion of elliptic space of three dimensions could be represented on a portion of a "hypersphere" in space of four dimensions, or the whole of elliptic space of three dimensions could be represented completely on a hypersphere, with the understanding that a point in elliptic space is represented by a pair of antipodal points on the hypersphere.

A hypersphere is a locus of constant curvature, just as a sphere is a surface of constant curvature. Analogy with the geometry of surfaces leads to the conception of the curvature of a three-dimensional locus in space of four dimensions, and just as the curvature of a surface can be determined at any point by intrinsic considerations, such as by measuring the angles of a geodesic triangle, so by similar measurements in the three-dimensional locus we could, without going outside that locus, obtain a notion of its curvature.

### 3. Application of differential geometry.

This was the path traversed by RIEMANN in his celebrated Dissertation. Space, he teaches us, is an example of a "manifold" of three dimensions, distinguished from other manifolds by nature of its homogeneity and the possibility of measurement. Space is unbounded, but not necessarily infinite. Thereby he expresses the possibility that the straight line may be of finite length, though without end—a conception that was absent from the minds of any of his predecessors. The position of a point $P$ can be determined by three numbers or coordinates, $x$, $y$, $z$; and if $x+dx$, $y+dy$, $z+dz$ are the values of the coordinates for a neighbouring point $Q$, then the length of the small element of length $PQ$, $=ds$, must be expressed in terms of the increments $dx$, $dy$, $dz$. If the increments are all increased in the same ratio, $ds$ will be increased in the same ratio, and if all the increments are changed in sign the value of $ds$ will be unaltered. Hence $ds$ must be an even root, square, fourth, etc., of a positive homogeneous function of $dx$, $dy$, $dz$ of the second, fourth, etc., degree. The simplest hypothesis is that $ds^2$ is a homogeneous function of $dx$, $dy$, $dz$ of the second degree, or by proper choice of coordinates $ds^2=$ a homogeneous linear expression in $dx^2$, $dy^2$, $dz^2$. For example, with rectangular coordinates in ordinary space, $ds^2=dx^2+dy^2+dz^2$.

By taking the analogy of Gauss' formulae for the curvature of a surface, Riemann defines a certain function of the differentials as the measure of curvature of the manifold. In order that congruence of figures may be possible, it is necessary that the measure of curvature should be everywhere the same; but it may be positive or zero. (Riemann had no conception of Lobachevsky's geometry, for which the measure of curvature is negative.) He gives without proof the following expression for the line-element. If $\alpha$ denotes the measure of curvature, then

$$ds = \sqrt{\Sigma dx^2}/(1 + \tfrac{1}{4}\alpha\Sigma x^2).$$

(Cf. Chap. V. § 35.) If $k$ is what has already been called the space-constant, $\alpha = 1/k^2$.

## 4. Free mobility of rigid bodies.

About the same time that Riemann's Dissertation was being published, Hermann von HELMHOLTZ (1821-1894) was conducting very similar investigations from the point of view of the general intuition of space, being incited thereto by his interest in the physiological problem of the localisation of objects in the field of vision.

Helmholtz [1] starts from the idea of congruence, and, by assuming certain principles such as that of *free mobility of rigid bodies*, and *monodromy*, *i.e.* that a body returns unchanged to its original position after rotation about an axis, he proves—what is arbitrary in Riemann's investigation—that the square of the line-element is a homogeneous function of the second degree in the differentials.

That the form of the function which expresses the distance between two points is limited by the possibility of the existence of congruent figures in different positions is shown as follows. Suppose we have five points in space, $A, B, C, D, E$. The position of each point is determined by three coordinates, and connecting each pair of points there is a certain expression involving the coordinates, which corresponds to the distance between the two points. Let

[1] "Ueber die Thatsachen, die der Geometrie zum Grunde liegen," *Göttinger Nachrichten*, 1868. An abstract of this paper was published in 1866.

us try to construct a figure $A'B'C'D'E'$ with exactly the same distances between pairs of corresponding points as the figure $ABCDE$. $A'$ may be taken arbitrarily. Then $B'$ must lie on a certain surface, since its coordinates are connected by one equation. $C'$ has to satisfy two conditions, and therefore lies on some curve, and then $D'$ is completely determined by its distances from $A'$, $B'$ and $C'$. Similarly $E'$ will be completely determined by its distances from $A'$, $B'$ and $C'$, but we cannot now guarantee that the distance $D'E'$ will be equal to $DE$. The distance-function is thus limited by one condition. And with more than five points a still greater number of conditions must be satisfied.[1]

It is customary to speak, as Helmholtz does, of the transformation of a figure into another congruent figure as a *displacement* of a single *rigid* figure from one position to another. This language often enables us to abbreviate our statements.

Thus, employing this language, we may argue for the general case as follows. If there are $n$ points, the figure has $3n$ degrees of freedom, and there are $\frac{1}{2}n(n-1)$ equations connecting the distances of pairs of points. But a rigid body has only 6 degrees of freedom; therefore the number of equations determining the distance-function is $\frac{1}{2}n(n-1)-3n+6=\frac{1}{2}(n-3)(n-4)$.

But it is necessary to avoid here a dangerous confusion. Points in space are fixed objects and cannot be conceived as altering their positions. When we speak of a motion of a rigid figure we are thinking of material bodies. The assumption which Helmholtz makes, which is expressed by the phrase, the "free mobility of rigid bodies," is thus

[1] This method was employed by J. M. de Tilly, *Bruxelles, Mém. Acad. Roy.* (8vo collection), **47** (1893), to find the expression for the distance-function without using infinitesimals.

simply an assumption that there is such a thing as absolute space.

While, psychologically, the idea of congruence may be based on the idea of rigid bodies, if it were really dependent upon the actual existence of rigid bodies it would have a very insecure foundation. Not only are the most solid bodies within our experience elastic and deformable, but modern researches in physics have given a high degree of probability to the conception that all bodies suffer a change in their dimensions when they are in motion relative to the aether. As all bodies, including our measuring rods, suffer equally in this distortion, however, we can never be conscious of it.

**5. Continuous groups of transformations.**

Helmholtz's researches, though of great importance in the history of the foundations of geometry, lacked the thoroughness which we would have expected had the author been a mathematician by profession.

The whole question was considered over again from a severely mathematical point of view by Sophus LIE[1] (1842-1899), who reduced the idea of motions to transformations between systems of coordinates, and congruence to invariance under such transformations. The underlying idea is that of a *group of transformations*.

Suppose we have a set of operations $R$, $S$, $T$, ... such that (1) the operation $R$ followed by the operation $S$ is again an operation (denoted by the product $RS$) of the set, and (2) $(RS)T = R(ST)$, then the set of operations is said to form a *group*. The operation, if it exists, which leaves the operand

[1] S. Lie, *Theorie der Transformationsgruppen*, vol. iii. (Leipzig, 1893), Abt. V. Kap. 20-24; and "Über die Grundlagen der Geometrie," *Leipziger Berichte*, **42** (1890).

unaltered, is called the *identical transformation*, and is denoted by 1.

Thus, if $R$, $S$, $T$ are the operations of rotation about a fixed point through 1, 2 and 3 right angles, the operations 1, $R$, $S$, $T$ form a group, and this is a *sub-group* of the group consisting of the 8 operations of rotation through every multiple of $\frac{\pi}{4}$.

The transformations which Lie considers are *infinitesimal transformations*, and the groups are *continuous groups*, such as the group of *all* the rotations about a fixed point. All the transformations which change points into points, straight lines into straight lines, and planes into planes form a continuous group which is called the general projective group.

The assumption from which Lie starts in his geometrical investigation is the "axiom of free mobility in the infinitesimal":

"If, at least within a certain region, a point $P$ and a line-element through $P$ are fixed, continuous motion is still possible, but if, in addition, a plane-element through $P$ is fixed, no motion is possible."

Starting then with the group of projective transformations, he determines the character of the transformations so that this assumption may be verified, and he proves that they form a group which leaves unaltered either a non-ruled surface of the second degree (real or imaginary ellipsoid, hyperboloid of two sheets or elliptic paraboloid), or a plane and an imaginary conic lying on this plane. This invariant figure is just the *Absolute*. The motions of space, therefore, form a sub-group of the general projective group of point-transformations which leave the Absolute invariant. And so, without Helmholtz's axiom of monodromy, but using a definite assumption of free mobility, Lie establishes that the only possible types of metrical geometry are the three in which the absolute is a real non-ruled quadric (hyperbolic geometry), an imaginary quadric (elliptic geometry), and a plane with an imaginary conic (euclidean geometry).

## 6. Assumption of coordinates.

There are several points on which the investigations of Riemann, Helmholtz and Lie admit of criticism. The outstanding difficulty which strikes one at once lies in the use of coordinates. How can we define the coordinates of a point before we have fixed the idea of congruence? This question has been settled by an appeal to the famous procedure of VON STAUDT (1798-1867), the founder of projective geometry. He has shown[1] how, by means of repeated application of the quadrilateral-construction for a harmonic range (see Chap. III. § 5), numbers may be assigned to all the points of a line. This, and other questions involved, have now been solved by the modern procedure of Pasch, Hilbert and the Italian school represented by Pieri. This procedure, which marks a return to the classical method of Euclid, consists in developing geometry as a purely logical system deduced from an appropriately chosen system of axioms or assumptions.

## 7. Space-curvature and the fourth dimension.

A misunderstanding, which is especially common among philosophers, has grown around Riemann's use of the term "curvature." Helmholtz, whose philosophical

[1] G. K. Ch. v. Staudt, *Geometrie der Lage*, Nürnberg, 1847, and *Beiträge zur Geometrie der Lage*, Nürnberg, 1856-57-60.

writings[1] are much better known than his mathematical researches, has unfortunately contributed largely to this error. The use of the term "space-curvature" has led to the idea that non-euclidean geometry of three dimensions necessarily implies space of four dimensions, for curvature of space has no meaning except in relation to a fourth dimension. But when we assert that space has only three dimensions, we thereby deny that space has four dimensions. The geometry of this space of three dimensions, whether it is euclidean or non-euclidean, follows logically from certain assumed premises, one of which will certainly be equivalent to the statement that space has not more than three dimensions (cf. Chap. II. § 14, footnote). The origin of the fallacy lies in the failure to recognise that the geometry on a curved surface is nothing but a representation of the non-euclidean geometry.

This is brought out still more clearly by the fact that, as non-euclidean geometry, elliptic or hyperbolic, can be represented on certain curved surfaces in euclidean space, the converse is also true, that euclidean geometry can be represented on certain curved surfaces in elliptic or hyperbolic space; and, of course, we do not consider the euclidean plane as being a curved surface.

While, therefore, the conception of non-euclidean space of three dimensions in no way implies necessarily space-curvature or a fourth dimension, it is still an interesting speculation to suppose that we exist really in a space of four dimensions, but with our experience confined to a certain curved locus in this space, just as Helmholtz's "two-dimensional beings" were confined to the surface

[1] H. v. Helmholtz, "The origin and meaning of geometrical axioms," *Mind*, **1** (1876), **3** (1878); also in *Popular Scientific Lectures* (London, 1881), vol. ii.

of a sphere in space of three dimensions, and acquired in this way the idea that their geometry is non-euclidean.

W. K. Clifford [1] has gone further than this and imagined that the phenomena of electricity, etc., might be explained by periodic variations in the curvature of space. But we cannot now say that this three-dimensional universe in which we have our experience is *space* in the old sense, for space, as distinct from matter, consists of a changeless set of terms in changeless relations. There are two alternatives. We must either conceive that space is really of four dimensions and our universe is an extended sheet of matter existing in this space, the aether [2] if we like; and then, just as a plane surface is to our three-dimensional intelligence a pure abstraction, so our whole universe will become an ideal abstraction existing only in a mind that perceives space of four dimensions—an argument which has been brought to the support of Bishop Berkeley! [3] Or, we must resist our innate tendencies to separate out space and bodies as distinct entities, and attempt to build up a monistic theory of the physical world in terms of a single set of entities, material points, conceived as altering their relations with time.[4] In either case it is not space that is altering its qualities, but matter which is changing its form or relations with time.

[1] *The Common Sense of the Exact Sciences* (London, 1885), chap. iv. § 19.

[2] Cf. W. W. Rouse Ball, " A hypothesis relating to the nature of the ether and gravity," *Messenger of Math.*, **21** (1891).

[3] See C. H. Hinton, *Scientific Romances*, First Series, p. 31 (London, 1886). For other four-dimensional theories of physical phenomena see Hinton, *The Fourth Dimension* (London, 1904).

[4] Cf. A. N. Whitehead, " On mathematical concepts of the material world," *Phil. Trans.*, A **205** (1906).

### 8. Proof of the consistency of non-euclidean geometry.

The characteristic feature of the second period in the history of non-euclidean geometry is brought out for the first time by BELTRAMI[1] (1835-1900), who showed that Lobachevsky's geometry is represented upon a surface of constant curvature. This is historically the first euclidean representation of non-euclidean geometry, and is of importance in providing a proof of the consistency of the non-euclidean systems. While the development of hyperbolic geometry in the hands of Lobachevsky and Bolyai led to no apparent internal contradiction, a doubt remained that inconsistencies might yet be discovered if the investigations were pushed far enough. This doubt was removed by Beltrami's concrete representation by means of the pseudosphere, which reduced the consistency of non-euclidean geometry to depend upon that of euclidean geometry, which everyone admits to be self-consistent.

Any concrete representation of non-euclidean geometry in euclidean space can be applied with the same object. In fact, the Cayley representation is more suitable for this purpose, since it affords an equally good representation of three-dimensional geometry. The advantage of Beltrami's representation is that distances and angles are truly represented, and the arbitrariness which may perhaps be felt in the logarithmic expressions for distances and angles is eliminated.

At the present time no *absolute* test of consistency is

[1] E. Beltrami, *Saggio di interpretazione della geometria non-euclidea*, Naples, 1868. Beltrami also showed that, since the equation of a geodesic in geodesic coordinates is linear, the surface can be represented on a plane, geodesics being represented by straight lines, and real points being represented by points lying within a fixed circle. He thus gave the transition from the geodesic to the projective representation of Cayley.

known to exist, and the only test which we can apply is to construct a concrete representation by means of a body of propositions whose consistency is universally granted. In the case of non-euclidean geometry the test which has just been applied suffices to prove the impossibility of demonstrating Euclid's postulate. For, if Euclid's postulate could be mathematically or logically proved, this would establish an inconsistency in the non-euclidean systems; but any such inconsistency would appear again in the concrete representation. The mathematical truth of the euclidean and the non-euclidean geometries is equally strong.

### 9. Which is the true geometry?

There being no *a priori* means of deciding from the mathematical or logical side which of the three forms of geometry does in actual fact represent the true relations of things, three questions arise:

(1) Can the question of the true geometry be decided *a posteriori*, or experimentally?

(2) Can it be decided on philosophical grounds?

(3) Is it, after all, a proper question to ask, one to which an answer can be expected?

### 10. Attempts to determine the space-constant by astronomical measurements.

Let us consider what form of experiment we can contrive to determine, if possible, the geometrical character of space. Essentially it must consist in the measurements of distances and angles, the sort of triangulation which is employed to determine the figure of the earth, but on a prodigiously larger scale. If we could measure the angles of some very large triangle, the difference between their sum and two right angles *might* give us the necessary data for determining the value of the space-constant. We do not say that such an experiment will give us the necessary data, for, as we shall see presently, the whole argument is destroyed by a vicious circle (§ 12); but let us assume, for the sake of illustrating the argument, that the experiment can be made, and see to what conclusions it leads.

The largest triangles, whose vertices are all accessible and whose angles we can measure directly, are far too small to allow of any discrepancy being observed. We must turn to astronomy to provide us with triangles of a suitable size. The largest triangles, of which two vertices are accessible, are those determined by a star and the observer in two different positions.

Let $S$ be a star and $E_1$, $E_2$ two positions of the earth at opposite ends of a diameter of its orbit, $C$ the sun; and

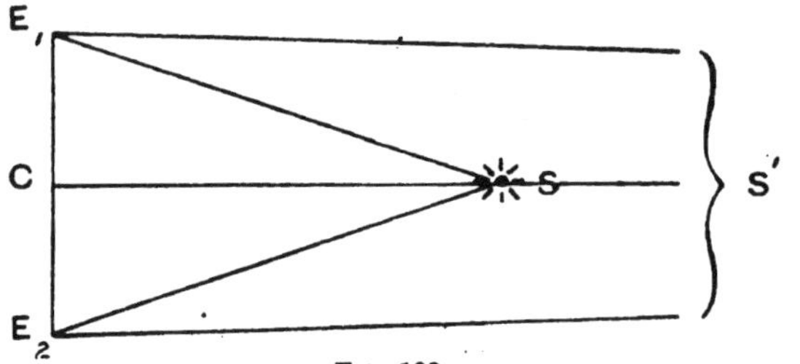


FIG. 103.

let $CS \perp E_1E_2$. The angle $E_1SC$, subtended by the earth's radius $R$ at the star, is called the *parallax* of the star; knowing this angle and applying euclidean geometry, we can find the star's distance.

There are two methods of determining the angle $E_1SC$. The first, or direct method, is to measure the angle $SE_1C$ by the transit circle. Then the parallax is, by assumption of euclidean geometry, the angle $\frac{\pi}{2} - SE_1C$. The second method, that of Bessel, is to compare the position of the star $S$ with those of neighbouring stars which, from their faintness and other considerations, are believed to be much farther away than $S$. Considering $S'$ as at infinity, and again assuming euclidean geometry, $E_1S' \parallel CS$ and the parallactic angle $E_1SC = S'E_1S$.

But on the hypothesis that geometry is hyperbolic, these two methods will give different results, and the angle $SE_1C + S'E_1S$ is in fact not equal to $\frac{\pi}{2}$, but is the parallel-angle corresponding to the distance $R$. Let $2\theta$ be the small difference $\left(\frac{\pi}{2} - SE_1C\right) - S'E_1S$; then

$$2\theta = \frac{\pi}{2} - \Pi(R).$$

Also

$$e^{-\frac{R}{k}} = \tan \tfrac{1}{2}\Pi(R) = \tan\left(\frac{\pi}{4} - \theta\right) = (1 - \tan\theta)/(1 + \tan\theta);$$

therefore

$$R/k = \log_e \{(1 + \tan\theta)/(1 - \tan\theta)\} = 2 \tan\theta, \text{ approx.}$$

Now we have records of the determination of the star $\alpha$ Centauri by both methods. An early measurement by the direct method yielded the value 1·14″, while Bessel's method gives the value 0·76″ ± 0·01″. Taking $2\theta$ therefore equal to 0·38″, we have $\tan\theta = 92 \times 10^{-8}$, and $k/R = 550000$ approx. The direct method is not susceptible of very great accuracy, and the value 1·14″ for the parallax is probably much too large, but even from these data, if we admit the soundness of our argument, we should be warranted in stating that the space-constant must be at least half a million times the radius of the earth's orbit.

The data, so far as they go, seem to favour the hypothesis of hyperbolic geometry rather than that of elliptic, since the calculation leads to a real value for $k$.

The hypothesis of elliptic geometry, however, leads to the result that a star would be visible in opposite directions unless there is some absorption of light in space.[1] If we assume that the light from a star which is at a distance of $\frac{1}{2}\pi k$ (*i.e.* half the total length of the straight line in elliptic space) is so diminished by absorption that the star becomes invisible, then the parallax of the farthest visible stars, measured by the direct method, would, as on the euclidean hypothesis without absorption, be zero. And if the light is totally absorbed in a distance of say $\frac{1}{4}\pi k$, the case would be similar to that on the hyperbolic hypothesis, or on the assumption of absorption in euclidean space. Thus, if we admit the hypothesis of absorption of light in free space, it becomes impossible to draw any definite conclusion as to the nature of actual space, except perhaps that the space-constant is very large.

The direct appeal to experiment therefore leads only to the conclusion that the space-constant, if not infinite, must be very large compared with any of the usual units of length, and is very large in comparison with the distances which we have ordinarily to deal with. These experiments do not contradict euclidean geometry, but they only verify it within the limits of experimental error. No amount of

[1] A complication, however, arises owing to the finite rate of propagation of light. The two images of the star seen in opposite directions will represent the star at different times, and in general therefore in different positions, so that, even if there were no absorption of light, the appearance of the sky would not necessarily be symmetrical. (Cf. W. B. Frankland, *Math. Gazette*, July 1913.)

experimental evidence of this kind can ever prove that the geometry of space is strictly euclidean, for there will always be a margin of error. On the other hand, so far as we have gone, it remains conceivable that further refinements in our instruments and more accurate information regarding the laws of absorption of light might enable us to establish an *upper* limit to the value of the space-constant, and thus demonstrate that the geometry of actual space is non-euclidean.

**11. Philosophy of space.**

This way of regarding experience as the source of our spatial ideas is in striking contrast to Kant's attitude towards space, which is expressed by his dicta: that space is not an empirical concept derived from external experience, but a framework already existing in the mind without which no external phenomena would be possible.[1] From the new point of view, geometry applied to actual space has become an experimental science, or a branch of applied mathematics. We are not forced to accept its axioms, but shall only do so when we find them convenient and in sufficiently close agreement with the facts of experience. Since Kant's time the intuitive has become discredited. We now know that there are things which formerly appeared to be intuitive which are in fact false. Thus, it was formerly believed that every continuous function possessed a differential coefficient; the proposition appeared, indeed, to be intuitive. But Weierstrass gave an example which showed that the belief was false. In the extreme empiricist view the parallel-postulate has to be ranked with the law of gravitation as a law of observation, which is verified within the limits of experimental error.

[1] I. Kant, *Critique of Pure Reason*, chap. i.

As regards the second question, therefore, the powers of philosophy have been narrowly circumscribed by the stricture laid upon intuition. Obviously the fact that a coherent mental picture can be formed of euclidean space does not constitute a proof that this is the form of actual space, since the same thing applies to the non-euclidean systems. But the philosopher may say he has an intuition of euclidean space. What does this mean? Has he an intuition that the sum of the angles of a triangle is equal to two right angles? Does he perceive intuitively that two straight lines which are both perpendicular to a third remain equidistant? What intuitions or beliefs would the philosopher have had if he had been deprived of powers of locomotion and the sense of touch, and been provided with only one eye? He would believe, because his eye told him so, that two railway lines converge to a point, that objects change their shapes when they are moved about; and he would perhaps demonstrate that the sum of the angles of a triangle is greater than two right angles. His intuitions are merely beliefs, and perhaps not even true ones.

We have really to distinguish between different kinds of space. The space of experience is brought to our knowledge through the senses principally of sight and touch, and is a composite of two spaces, "visual space" and "tactual space." Pure visual space, which is the limited field of our imaginary one-eyed sessile philosopher, is a crude elliptic two-dimensional space;[1] the three-dimensional form of tactual space is conditioned probably in part by the semi-circular canals of the ear. From this composite

[1] Cf. Thomas Reid, *An Inquiry into the Human Mind*, Edinburgh, 1764, chap. vi. "On Seeing," § 9 "Of the geometry of visibles."

space, which is far from being the beautiful mathematical continuum which we have arrived at after generations of thought, we get by abstraction a conceptual space which is conditioned only by the laws of logic, but to which we find it convenient to ascribe the particular form which we call euclidean space, for the reason that this is the simplest of the logically possible forms which correspond with sufficient closeness to the space of experience. Whether there is, besides these, an intuitional space, we shall leave to philosophy to settle if it can. We may, perhaps, leave Kant in possession of an *a priori* space as the framework of his external intuitions, but this space is amorphous, and only experience can lead us to a conception of its geometrical properties.

**12. The inextricable entanglement of space and matter.**

A further point—and this is the " vicious circle " of which we spoke above—arises in connection with the astronomical attempts to determine the nature of space. These experiments are based upon the received laws of astronomy and optics, which are themselves based upon the euclidean assumption. It might well happen, then, that a discrepancy observed in the sum of the angles of a triangle could admit of an explanation by some modification of these laws, or that even the absence of any such discrepancy might still be compatible with the assumptions of non-euclidean geometry.

" All measurement involves both physical and geometrical assumptions, and the two things, space and matter, are not given separately, but analysed out of a common experience. Subject to the general condition that space is to be changeless and matter to move about in space, we can explain the same observed results in many different ways by making compensatory changes in the qualities that we assign to space and the qualities we assign to matter. Hence it seems theoretically impossible to decide by any experiment what are the qualities of one of them in distinction from the other." [1]

It was on such grounds that Poincaré [2] maintained the essential impropriety of the question, "Which is the true geometry?" In his view it is merely a matter of convenience. Facts are and always will be most simply described on the euclidean hypothesis, but they can still be described on the non-euclidean hypothesis, with suitable modifications of the physical laws. To ask which is the true geometry is then just as unmeaning as to ask whether the old or the metric system is the true one. The conclusion thus arrived at by Poincaré is quite akin to the modern doctrine in physics expressed by the Principle of Relativity. Just as, according to this doctrine, it is impossible by any means to obtain a knowledge of absolute motion, so, according to Poincaré, it is beyond our power to obtain a knowledge of absolute space.

[1] Mr. C. D. Broad, with whom I have discussed this chapter, has put this point of view so well that I quote his words.

[2] H. Poincaré, *La science et l'hypothèse* (Paris, 1902), chap. v.; English translation by W. J. Greenstreet, London, 1905.

## CHAPTER VII.

### RADICAL AXES, HOMOTHETIC CENTRES AND SYSTEMS OF CIRCLES.

**1. Common points and tangents to two circles.**

Two circles intersect in four points and have four common tangents. Various cases arise according as these points and lines are coincident or imaginary in pairs.

In hyperbolic geometry two equidistant-curves whose axes intersect have their common points and tangents all real. A proper circle which cuts both branches of an equidistant-curve has four real common tangents with it. If it cuts only one branch, two of the common points and two of the common tangents are imaginary. Two proper circles cannot have more than two of their common points real; their common tangents are then two real and two imaginary. If two proper circles do not intersect, their common tangents are all real or all imaginary. The case of four real common points and four imaginary common tangents cannot occur in hyperbolic geometry; two real and two imaginary common points can only occur along with two real and two imaginary common tangents.

In elliptic geometry, if two circles intersect in two real and two imaginary points, they have two real and two imaginary common tangents. If each lies entirely outside the other, their common points are all imaginary and their common tangents are all real. The absolute polars of two such circles have four real common points and their common tangents all imaginary. If one lies entirely within the other, their common points and tangents are all imaginary. The case of four real common points and four real common tangents cannot occur in elliptic geometry.

## 2. The power of a point with respect to a circle.

Let $O$ be a fixed point in the plane of a proper circle with centre $C$ and radius $a$. Through $O$ draw any secant cutting

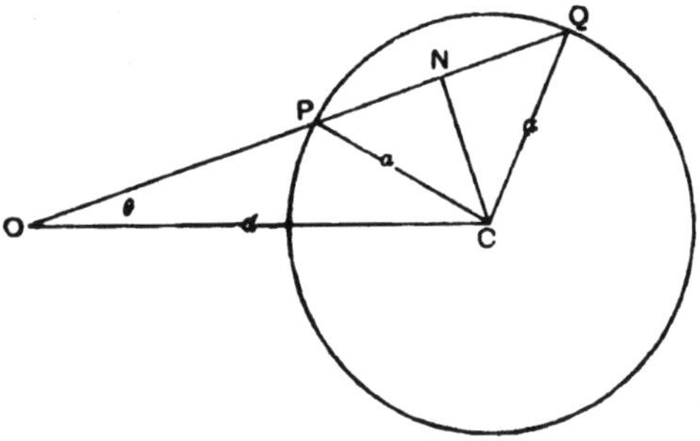


FIG. 104.

the circle in $P$, $Q$. Draw $CN \perp OPQ$. Let $OC = d$ and $COP = \theta$, $OP = r$, $OQ = r'$, so that

$$ON = \tfrac{1}{2}(r' + r), \; PN = \tfrac{1}{2}(r' - r).$$

Now[1]

$$\cos d = \cos CN \cos \tfrac{1}{2}(r' + r),$$

$$\cos a = \cos CN \cos \tfrac{1}{2}(r' - r).$$

Therefore
$$\frac{\cos d}{\cos a} = \frac{\cos \frac{1}{2}(r' + r)}{\cos \frac{1}{2}(r' - r)} = \frac{1 - \tan \frac{1}{2}r \tan \frac{1}{2}r'}{1 + \tan \frac{1}{2}r \tan \frac{1}{2}r'};$$

therefore $\tan \frac{1}{2}r \tan \frac{1}{2}r' = \text{const.} = \tan \frac{1}{2}(d + a) \tan \frac{1}{2}(d - a)$.

[1] Elliptic geometry is taken as the standard case, and the space-constant $k$ is taken as the unit of length.

In hyperbolic geometry tan is replaced by $i$ tanh. This product may be called the *power* of the point $O$ with respect to the circle. It is positive if $O$ is outside, negative if $O$ is inside the circle. In the former case, if $t$ is the length of the tangent from $O$ to the circle, the power of $O$ is equal to $\tan^2\frac{1}{2}t$.

**3. Power of a point with respect to an equidistant-curve.**

(1) Let the secant cut one branch of the curve in $P$, $Q$, *i.e.* in hyperbolic geometry the secant does not cut the

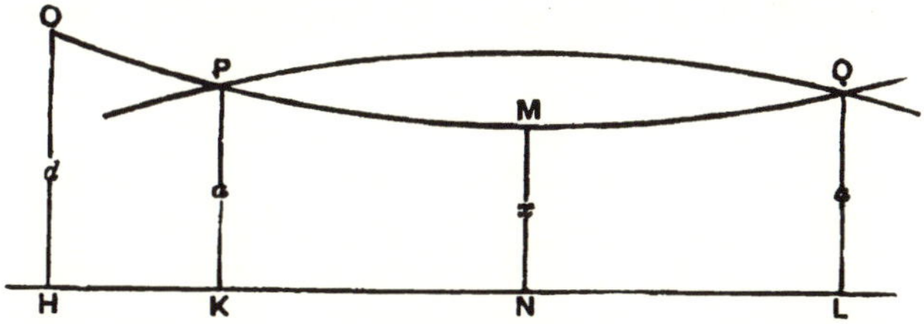


FIG. 105.

axis of the curve, in elliptic geometry neither of the finite segments $OP$, $OQ$ cuts the axis.

Let $M$ be the middle point of $PQ$, and draw $MN \perp$ the axis; then $MN$ is also $\perp PQ$. Draw $OH \perp$ the axis. Let $OH = d$, $MN = x$, $OP = r$, $OQ = r'$, so that $OM = \frac{1}{2}(r' + r)$, $PM = \frac{1}{2}(r' - r)$.

Then, from the trirectangular quadrilaterals $OHNM$, $PKNM$,

$$\sin d = \cos \tfrac{1}{2}(r' + r) \sin x, \quad \sin a = \cos \tfrac{1}{2}(r' - r) \sin x;$$

therefore $$\frac{\sin d}{\sin a} = \frac{\cos \frac{1}{2}(r' + r)}{\cos \frac{1}{2}(r' - r)} = \frac{1 - \tan \frac{1}{2}r \tan \frac{1}{2}r'}{1 + \tan \frac{1}{2}r \tan \frac{1}{2}r'};$$

therefore $\tan \frac{1}{2}r \tan \frac{1}{2}r' = \text{const.} = \tan \frac{1}{2}(a - d)/\tan \frac{1}{2}(d + a)$.

(2) Let the secant cut both branches of the curve, *i.e.* the point of intersection $A$ with the axis is real and one of the segments $OP$, $OQ$ cuts the axis.

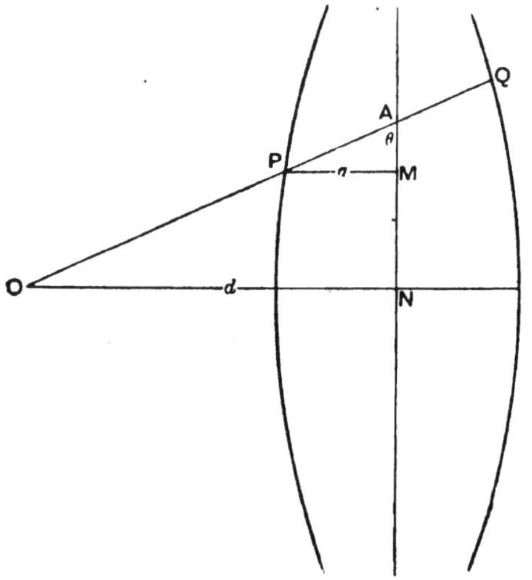


FIG. 106.

Let $OAN = \theta$, $ON = d$, $OP = r$, $OQ = r'$, so that

$$OA = \tfrac{1}{2}(r' + r), \quad PA = \tfrac{1}{2}(r' - r).$$

Then
$$\sin d = \sin \tfrac{1}{2}(r' + r) \sin \theta,$$
$$\sin a = \sin \tfrac{1}{2}(r' - r) \sin \theta.$$

Therefore
$$\frac{\sin d}{\sin a} = \frac{\sin \frac{1}{2}(r' + r)}{\sin \frac{1}{2}(r' - r)} = \frac{\tan \frac{1}{2}r' + \tan \frac{1}{2}r}{\tan \frac{1}{2}r' - \tan \frac{1}{2}r}.$$

Therefore
$$\frac{\tan \frac{1}{2}r}{\tan \frac{1}{2}r'} = \text{const.} = \frac{\tan \frac{1}{2}(d - a)}{\tan \frac{1}{2}(d + a)}.$$

*Note.* Figs. 105 and 106 have been drawn for the case of hyperbolic geometry. In elliptic geometry the equidistant-curve is convex towards the axis. In Fig. 105, in this case, either $OH < PK$ or $O$ lies between $P$ and $Q$. If $O$ is the same point in the two figures, the values of $\tan \frac{1}{2}r / \tan \frac{1}{2}r'$ and $\tan \frac{1}{2}r \tan \frac{1}{2}r'$, respectively for the secant which cuts and the secant which does not cut the axis, are equal.

Hence, *if a variable line through a fixed point O cuts a circle in P, Q and its axis in A, either the ratio or the product of the tangents of half the segments OP, OQ is constant, according as* (1) *one, or* (2) *both or neither of the segments contains the point A. If OT is a tangent to the curve, the constant is equal to* $\tan^2 \frac{1}{2}OT$, *and is called the power of the point O with respect to the circle.*

The two cases are simply explained in elliptic geometry. Let the dotted circle $AA'$ represent the axis of the circle, which is

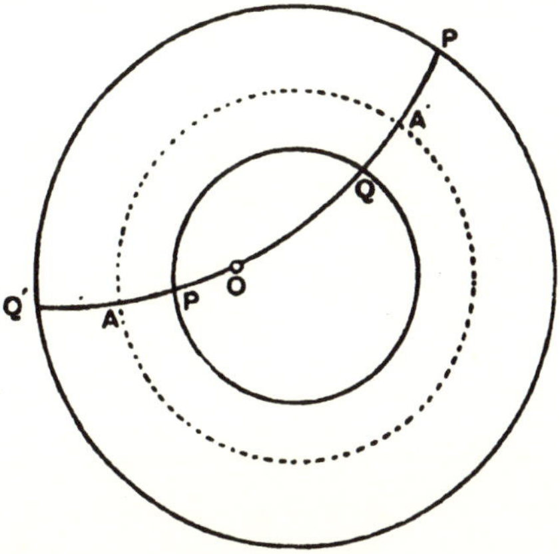


FIG. 107.

represented in the diagram by a pair of circles. The secant cuts the two circles in $P, P'$; $Q, Q'$; and the axis in $A, A'$. These pairs of course represent single points.

$$AA' = PP' = QQ' = \pi ;$$

therefore $$OQ' = \pi - OQ.$$

Therefore $$\tan \tfrac{1}{2}OP \tan \tfrac{1}{2}OQ = \tan \tfrac{1}{2}OP \cot \tfrac{1}{2}OQ' = \frac{\tan \frac{1}{2}OP}{\tan \frac{1}{2}OQ'}.$$

**4.** Reciprocally, if $P$ is a variable point on a fixed line $PN$,

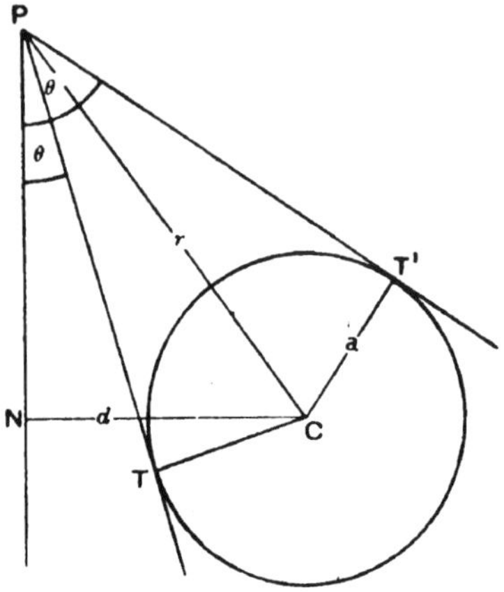


FIG. 108.

and the tangents $PT$, $PT'$ from $P$ to a fixed circle make angles $\theta$, $\theta'$ with $PN$, we have in Fig. 108,

$$\sin d = \sin r \sin \tfrac{1}{2}(\theta' + \theta),$$
$$\sin a = \sin r \sin \tfrac{1}{2}(\theta' - \theta),$$
$$\frac{\sin d}{\sin a} = \frac{\sin \frac{1}{2}(\theta' + \theta)}{\sin \frac{1}{2}(\theta' - \theta)} = \frac{\tan \frac{1}{2}\theta' + \tan \frac{1}{2}\theta}{\tan \frac{1}{2}\theta' - \tan \frac{1}{2}\theta};$$

whence $$\frac{\tan \frac{1}{2}\theta}{\tan \frac{1}{2}\theta'} = \text{const.} = \frac{\tan \frac{1}{2}(d - a)}{\tan \frac{1}{2}(d + a)}.$$

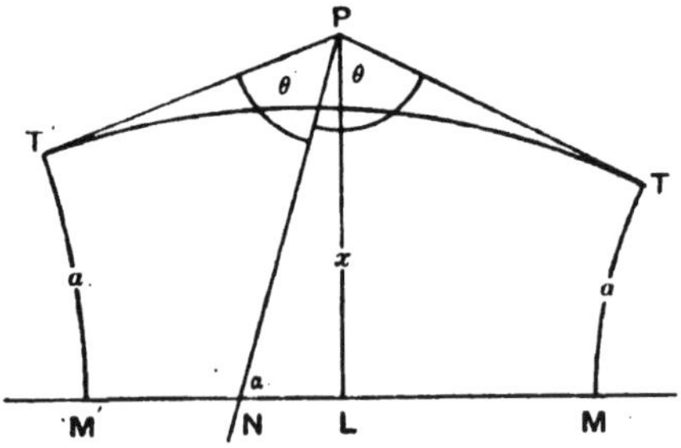


FIG. 109.

This result is true also in euclidean geometry, the constant reducing to $(d-a)/(d+a)$.

For an equidistant-curve, let the line cut the axis in $N$ at an angle $\alpha$ (Fig. 109).

Then, $\theta$ and $\theta'$ being taken positively,

$$TPL = T'PL = \tfrac{1}{2}(\theta + \theta'),$$

$$NPL = \tfrac{1}{2}(\theta - \theta'),$$

$$\cos \alpha = \cos x \sin \tfrac{1}{2}(\theta - \theta'),$$

$$\cos a = \cos x \sin \tfrac{1}{2}(\theta + \theta');$$

whence, as before, $\dfrac{\tan \frac{1}{2}\theta}{\tan \frac{1}{2}\theta'}$ is constant.

If the angles $\theta$, $\theta'$ are measured in the same sense, then for $\theta'$ we must put $\pi - \theta'$, and we have

$$\tan \tfrac{1}{2}\theta \tan \tfrac{1}{2}\theta' = \text{const.}$$

If, the angles $\theta$, $\theta'$ being measured in the same sense, both or neither of them contains the line joining $P$ to the centre, then we have (Fig. 110)

$$LPC = \tfrac{1}{2}(\theta' + \pi + \theta) = \frac{\pi}{2} + \tfrac{1}{2}(\theta' + \theta),$$

$$NPC = \frac{\pi}{2} - \tfrac{1}{2}(\theta' + \theta),$$

$$TPC = \tfrac{1}{2}TPT' = \tfrac{1}{2}(\pi - \theta' + \theta) = \frac{\pi}{2} - \tfrac{1}{2}(\theta' - \theta).$$

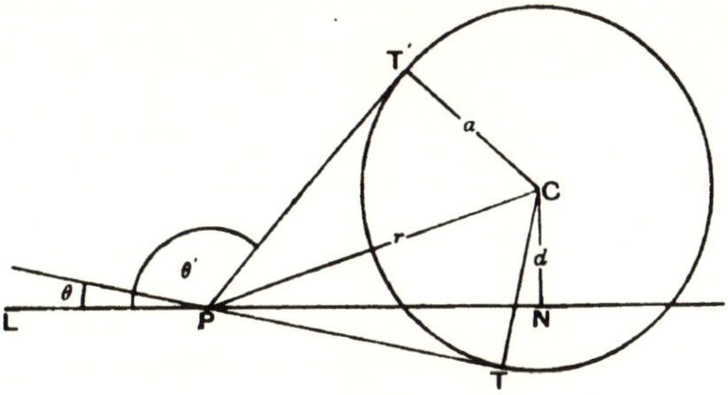


FIG. 110.

Then $$\sin d = \sin r \cos \tfrac{1}{2}(\theta' + \theta),$$
$$\sin a = \sin r \cos \tfrac{1}{2}(\theta' - \theta),$$
and $$\tan \tfrac{1}{2}\theta \tan \tfrac{1}{2}\theta' = \text{const.}$$

Hence, *if from a variable point on a fixed line $l$ the tangents to a circle are $p$, $q$, and the line to the centre is $a$, either the ratio or the product of the tangents of half the angles* $(lp)$, $(lq)$ *is constant, according as* (1) *one, or* (2) *both or neither of the angles contains the line $a$.*

### 5. Angles of intersection of two circles.

Since two circles may intersect in four points, there are four angles of intersection to consider.

It is easy to show geometrically that if two circles have only two real points of intersection, the two angles of intersection are equal.

Suppose a circle cuts an equidistant-curve in four points, $P$, $P'$ on one branch, $Q$, $Q'$ on the other branch. Then,

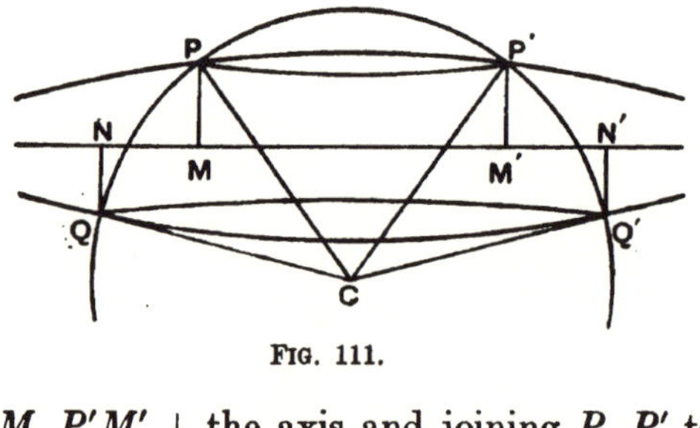


FIG. 111.

drawing $PM$, $P'M' \perp$ the axis and joining $P$, $P'$ to $C$, the centre of the circle,

$$\angle CPP' = CP'P, \quad \angle MPP' = M'P'P;$$

therefore $$\angle MPC = M'P'C,$$

and the angles of intersection at $P$, $P'$ are equal, and similarly the angles of intersection at $Q$, $Q'$ are equal.

But $\angle CPM + CQN = 2CPQ$, since $\angle MPQ = NQP$; therefore $CPM$ and $CQN$ are not in general equal. If $CPM = CQN$, then each $= CPQ = CQP$; $M$ and $N$ then coincide, and $C$ lies on the axis of the equidistant-curve.

Similarly, it may be shown that if two equidistant-curves intersect in four points, the angles at the points of intersection which are on different branches are equal, but all four angles cannot be equal unless the axes are at right angles.

When two of the angles of intersection are right, the circles are said to cut *orthogonally*. All four angles cannot be right, for then the centre $C$ of the one circle must lie on the axis of the other, and if $CT$, $CT'$ are the tangents to the second circle, $CT$ is a radius of the first circle. But $CT$ is a quadrant; therefore the first circle must reduce to two coincident straight lines.

### 6. Radical axes.

Let $P$, $P'$, $Q$, $Q'$ be the points of intersection of two circles, with axes $\alpha = 0$ and $\beta = 0$. Then, if $S = 0$ is the equation of the absolute, the equations of the circles can be written

$$S - \alpha^2 = 0, \quad S - \beta^2 = 0.$$

The equation $(S - \alpha^2) - (S - \beta^2) = 0$ represents a conic passing through their common points, but this breaks up into the two straight lines $\alpha \pm \beta = 0$, and these represent a pair of common chords which pass through the point of intersection of the axes. They form with $\alpha$ and $\beta$ a harmonic pencil.

If $\gamma$ is the polar of the intersection of the axes, *i.e.* the line of centres, the other pairs of common chords pass through $\alpha\gamma$ and $\beta\gamma$.

If we take any point $O$ on one of the first pair of common

chords, say $PP'$, the power of $O$ with respect to either circle is $\tan \frac{1}{2}OP/\tan \frac{1}{2}OP'$. These two lines are therefore the locus of points from which equal tangents can be drawn to the two circles.

But if we take a point $O$ on $PQ$, the power with respect to one circle is the product, and with respect to the other circle the ratio of $\tan \frac{1}{2}OP$ and $\tan \frac{1}{2}OQ$, and this chord does not possess the property of equal tangents.

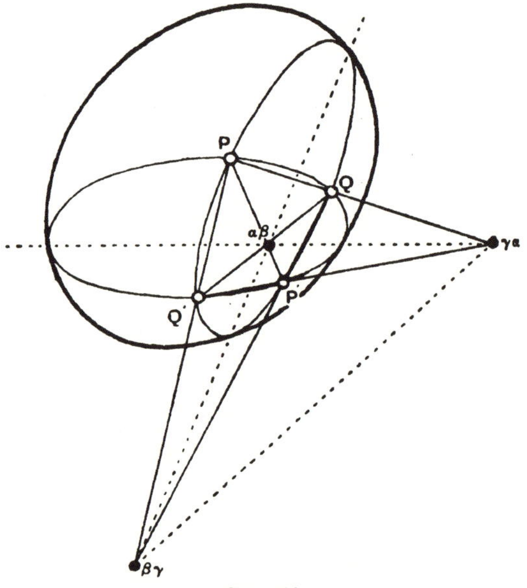


FIG. 112.

Hence, of the three pairs of common chords of two circles, one pair pass through the intersection of the axes and are harmonically separated by them, and possess the property that the tangents from any point on either to the two circles are equal.

These two lines are called the *radical axes* of the two circles.

### 7. Homothetic centres.

Reciprocally, two circles have four common tangents, which intersect in three pairs of points. One pair lie on the line joining the centres, and are harmonically separated by them, the other pairs lie on the lines joining the centres to the pole of the line of centres. The first pair possess the property that any line drawn through one of them cuts the two circles at equal angles. These two points are called the *homothetic centres* of the two circles.

### 8. Radical centres and homothetic axes.

*The three pairs of radical axes of three circles taken in pairs pass through four points, the radical centres of the three circles.*

Let $ABC$ be the triangle formed by the axes $a$, $b$, $c$ of the three circles; a pair of radical axes $\alpha_1\alpha_2$, $\beta_1\beta_2$, $\gamma_1\gamma_2$ passes through each of these points.

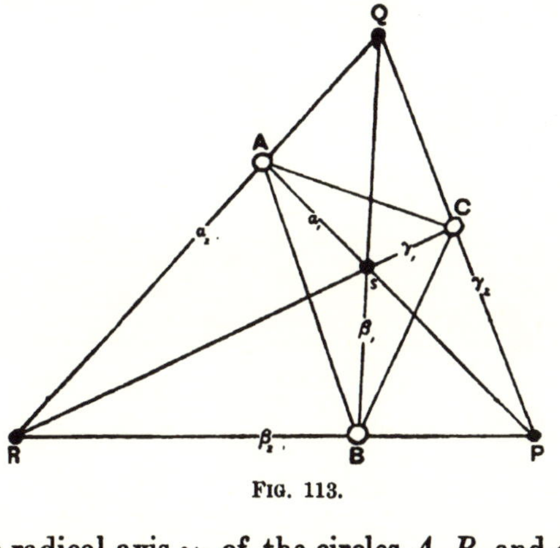


FIG. 113.

If one radical axis $\gamma_2$ of the circles $A$, $B$, and one radical axis $\beta_2$ of the circles $A$, $C$ intersect in $P$, then the tangents from $P$ to the three circles are all equal. Therefore $P$ lies on a radical axis $\alpha_1$ of the circles $B$, $C$. We have then $\alpha_1, \beta_2, \gamma_2$ concurrent in $P$. Let $\gamma_1$ cut $\alpha_1$ in $S$, and join $BS$. Then, since $(ab, \gamma_1\gamma_2)$ is harmonic, $B(AC, SP)$ is harmonic; therefore $BS$ is $\beta_1$, *i.e.* $\alpha_1, \beta_1, \gamma_1$ pass through $S$. Similarly $\alpha_2, \beta_1, \gamma_2$ are concurrent in $Q$, and $\alpha_2, \beta_2, \gamma_1$ in $R$. The quadrangle $PQRS$ has $ABC$ as harmonic triangle.

Reciprocally, *the three pairs of homothetic centres of three circles taken in pairs lie in sets of three on four lines, the homothetic axes of the three circles.* They form a complete quadrilateral, whose harmonic triangle is the triangle formed by the centres of the circles.

## 9. Coaxal circles in elliptic geometry.

The locus of the centre of a circle which passes through two fixed points $D_1$, $D_2$ on a line $l$ consists of the two perpendicular bisectors $OL$, $O'L$ of the segments $D_1D_2$ and $D_2D_1$ (Fig. 114). All the circles through $D_1$, $D_2$ therefore fall into two groups; any two circles belonging to the same group have $l$ as a radical axis. Each group is therefore called a system of *coaxal circles* with common points $D_1$, $D_2$. When the centre is at $O$, the circle is a minimum, and it increases up to a maximum, which is just the straight line $l$ itself, when the centre is at $L$.

Let $C_1$, $C_2$ on $OL \equiv l'$ be the centres of two circles of the one system, and take two points $K_1$, $K_2$ on $l$. Draw the tangents $K_1U_1$, $K_1U_2$, $K_2V_1$, $K_2V_2$ to the circles $C_1$, $C_2$. Then $K_1U_1 = K_1U_2$ and $K_2V_1 = K_2V_2$. Hence the points $U$ lie on a circle with centre $K_1$, and $V$ on a circle with centre $K_2$. Also, since $K_1U_1$ is a tangent to the circle $C_1$ and a radius of the circle $K_1$, $C_1U_1$ is a tangent to the circle $K_1$; and since $C_1U_1 = C_1V_1$, $C_1$ lies on a radical axis of the circles $K_1$, $K_2$. Hence the circles $K$ have $l'$ as a radical axis. We get then a system of coaxal circles $K$ associated with the system $C$, and every circle of the one system cuts orthogonally every circle of the other system. As $KD_1$ diminishes the circle tends to vanish. $D_1$, $D_2$ are called the *limiting points* of the $K$ system. If $K$ lies

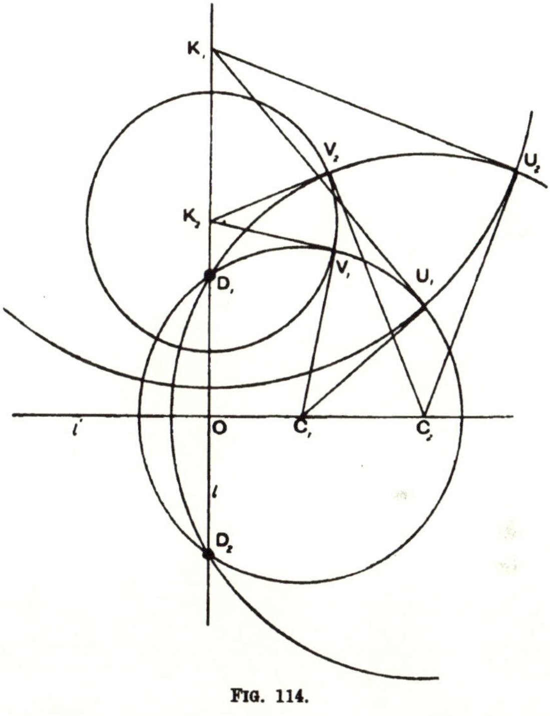


FIG. 114.

in the segment $D_1OD_2$, the circle is imaginary. As $K$ approaches $O'$, the circle becomes the straight line $l'$. The $K$ system is a non-intersecting system, *i.e.* it has imaginary common points. The $C$ system has imaginary limiting points.

If the segment $D_1D_2$ vanishes so that the common points

$D_1$, $D_2$ coincide at $O$, the circles $C$ all touch the line $l$ at $O$, and the circles $K$ all touch the line $l'$ at $O$.

If the segment $D_1D_2$ becomes $\pi$, so that the common points coincide at $O'$, the circles $C$ all reduce to straight lines passing through $O'$, while the circles $K$ become concentric circles with centre $O'$.

### 10. Homocentric circles.

The locus of the centre of a circle which touches two fixed lines $d_1$, $d_2$ through a point $L$ consists of the two bisectors $o$, $o'$ of the angles between $d_1$, $d_2$ (Fig. 115). All the circles touching $d_1$, $d_2$ therefore fall into two groups; any two circles belonging to the same group have $L$ as a homothetic centre. Each group is therefore called a system of *homocentric circles* with common tangents $d_1$, $d_2$. When the centre is at $L$, the circle is a minimum and reduces to the point $L$ itself; as the centre moves along $o'$, the circle increases up to a maximum when the centre is at $O$, the pole of $o$.

Let $c_1$, $c_2$ through $L'$, the intersection of $o$ and $l$, be the axes of two circles of the one system. Take two lines $k_1$, $k_2$ through $L$, and let $u_1$, $u_2$, $v_1$, $v_2$ be the tangents to the circles $C_1$, $C_2$ at their points of intersection with $k_1$, $k_2$. Then the angles $(k_1u_1)=(k_1u_2)$ and $(k_2v_1)=(k_2v_2)$. Hence the lines $u$ are tangents to a circle with axis $k_1$, and $v$ are tangents to a circle with axis $k_2$. Also, since $(k_1u_1)$ lies on the circle $C_1$ and on the axis of the circle $K_1$, and since the angle $(c_1u_1)$ $=(c_1v_1)$, $c_1$ passes through a homothetic centre of the circles $K_1$, $K_2$. Hence the two circles $K$ have $L'$ as a homothetic centre. We get then a system of homocentric circles $K$ associated with the system $C$, and every circle of the one system is tangentially distant a quadrant from every circle of the other system. As $k$ approaches $d_1$ the circle $K$ becomes the straight line $d_1$. $d_1, d_2$ are called the limiting lines of the $K$ system. If $k$ lies outside the angle $d_1d_2$,

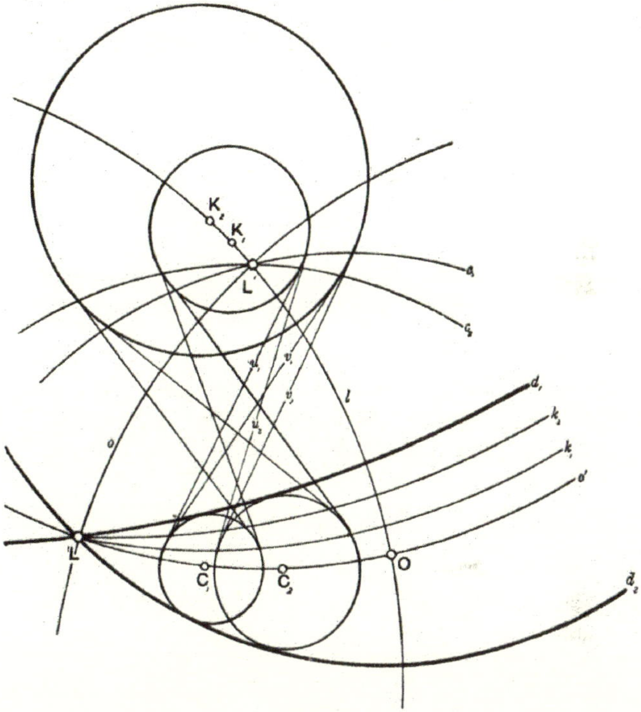

FIG. 115.

the circle becomes imaginary. As $k$ approaches $o'$, the circle reduces to the point $L'$. The $K$ system has imaginary common tangents; the $C$ system has imaginary limiting lines.

If the angle $d_1d_2$ vanishes, so that the common tangents

$d_1$, $d_2$ coincide with $o'$, the circles $C$ all reduce to points on $o'$, while the circles $K$ become concentric circles with axis $o'$.

If the angle $d_1 d_2$ becomes $\pi$, so that the common tangents $d_1$, $d_2$ coincide with $o$, the circles $C$ all touch $o$ at $L$, and the circles $K$ all touch $o$ at $L'$.

**11.** In euclidean and hyperbolic geometry this duality does not hold, since in euclidean geometry the envelope of a system of lines cutting a fixed line at a constant angle is a point at infinity, and in hyperbolic geometry it is an ideal circle. In hyperbolic geometry, as $K$ goes to infinity the circle becomes a horocycle. Between the horocycle and the straight line lies a system of branches of equidistant-curves. The other branches complicate the figure as they intersect the other circles of the system. The same thing, of course, occurs in the other coaxal system passing through $D_1$, $D_2$.

In euclidean geometry a system of coaxal circles is a linear system, *i.e.* through a given point only one circle of the system passes. In non-euclidean geometry, through three given points four circles pass, *i.e.* four circles can be drawn through any point $P$ to pass through two fixed points $X$, $Y$. Denote these circles by $PXY$, $P'XY$, etc.; then of the four circles, $PXY$, $P'XY$ have their centres on the one perpendicular bisector of $XY$, and belong to the one coaxal system, while $PX'Y$, $PXY'$ belong to the other. Hence, through a given point there pass only two circles of a given coaxal system.

In euclidean geometry a system of coaxal circles is equivalent to a system of conics through four points; in non-euclidean geometry it is equivalent to a system of conics through two points and having double contact with a fixed conic.[1] The reciprocal system, *i.e.* a system of circles touching two lines, is equivalent in non-euclidean geometry, to a system of conics touching two lines and having double contact with a fixed conic; and in euclidean geometry to a system of conics passing through two points and touching two lines. Hence the complexity of the latter system compared with a system of coaxal circles.

The analytical treatment of systems of coaxal circles in non-euclidean geometry can be reduced to the consideration of linear systems in the following way.

## 12. Linear equation of a circle.

If $(x, y, z)$ are the actual Weierstrass coordinates, the equation of a circle, with centre $(x_1, y_1, z_1)$ and radius $r$, is

$$xx_1 + yy_1 + k^2zz_1 = k^2 \cos\frac{r}{k}.$$

Let $a = \lambda x_1$, $b = \lambda y_1$, $c = \lambda k^2 z_1$, $d = -\lambda k^2 \cos\frac{r}{k}$, so that

$$k^2a^2 + k^2b^2 + c^2 = \lambda^2k^2(x_1^2 + y_1^2 + k^2z_1^2) = \lambda^2k^4 = p^2, \text{ say.}$$

Then the equation reduces to

$$ax + by + cz + d = 0.$$

The non-homogeneous linear equation, with real coefficients, in actual Weierstrass coordinates, therefore represents a circle with centre $(a, b, c/k^2)$, axis $ax + by + cz = 0$, and radius $r$, such that

$$\cos\frac{r}{k} = \text{the positive value of } \frac{d}{p},$$

where $p^2 = k^2a^2 + k^2b^2 + c^2$.

In elliptic geometry $k^2$ is positive and $p^2$ is always positive. The centre is always real, and the radius is real if $d^2 = k^2a^2 + k^2b^2 + c^2$. If $d = 0$ the circle becomes a straight line, and if $d = p$ it reduces to a point.

In hyperbolic geometry, changing the sign of $k^2$, we have

$$p^2 = c^2 - k^2a^2 - k^2b^2, \text{ and } \cosh\frac{r}{k} = \text{the positive value of } \frac{d}{p}.$$

[1] It is therefore exactly equivalent to a system of circles in euclidean geometry having double contact with a fixed conic. The limiting points are represented by the foci of the conic.

The centre is real, ideal or at infinity, according as $p^2 >, =$ or $<0$. The curve is therefore

A real circle if $d^2 > c^2 - k^2a^2 - k^2b^2 > 0$, reducing to a point if $d^2 = c^2 - k^2a^2 - k^2b^2 > 0$.

An imaginary circle if $c^2 - k^2a^2 - k^2b^2 > d^2 > 0$.

An equidistant-curve if $c^2 < k^2(a^2+b^2)$, reducing to a straight line if $d=0$.

A horocycle if $c^2 = k^2(a^2+b^2)$.

The two equations $ax+by+cz \pm d=0$ represent the same circle. In elliptic geometry this is verified, since $(x, y, z)$ and $(-x, -y, -z)$ represent the same point. In hyperbolic geometry, for a proper circle or a horocycle only one of these equations can be satisfied, since $z$ must be positive; for an equidistant-curve the two equations represent the two branches.

The points of intersection of two circles $a_1x+b_1y+c_1z \pm d_1=0$ and $a_2x+b_2y+c_2z \pm d_2=0$ are found by solving these equations simultaneously with the equation $x^2+y^2+k^2z^2=k^2$. These give four sets of values of $x^2$, $y^2$, $z^2$, and therefore four points of intersection.

## 13. Systems of circles.

If $S_1=0$ and $S_2=0$ are equations of circles in this form, $S_1+\lambda S_2=0$ represents a circle, and for all values of $\lambda$ represents a pencil of circles passing through two fixed points. If $d_1=d_2$, the circle $S_1-S_2=0$ of the system reduces to a straight line, and if $d_1+d_2=0$ the circle $S_1+S_2=0$ is another straight line. These are the radical axes of the two circles.

$S_1+\lambda S_2+\mu S_3=0$ represents a linear two-parameter system or bundle of circles. If a circle of the system passes through the point $x'$,

$$S_1'+\lambda S_2'+\mu S_3'=0,$$

and we have $$(S_1S_3'-S_1'S_3)+\lambda(S_2S_3'-S_2'S_3)=0,$$

which represents a linear one parameter system or pencil of circles. Hence all circles of a bundle which pass through one fixed point form a coaxal system and pass through another fixed point.

If $d_1+\lambda d_2+\mu d_3=0$, we get a pencil of straight lines If the vertex of this pencil is real, choose it as origin; then the linear system can be reduced to the form $x+\lambda y+\mu(z+c)=0$. Then one radical axis of every pair of circles of the system passes through the origin, *i.e.* the circles have a common radical centre at the origin. If tangents are drawn from this point to the circles of the system they are all equal, and hence all circles of a bundle cut orthogonally a fixed circle. In hyperbolic geometry the orthogonal circle may be a proper circle, real or imaginary, an equidistant-curve or a horocycle. If the orthogonal circle reduces to a point, all the circles pass through this point.

All these results admit, in the case of elliptic geometry, of a simple interpretation by means of the central projection of the sphere. To a plane corresponds a circle, to an axial pencil of planes corresponds a pencil of circles, and to a bundle of planes through a fixed point $O$ corresponds a bundle of circles, the orthogonal circle of which corresponds to the polar plane of $O$ with respect to the sphere.

This representation fails in hyperbolic geometry, since the sphere becomes imaginary, but there is a correspondence between the circles of the hyperbolic plane and the planes of hyperbolic space.

**14. Correspondence between circles and planes in hyperbolic geometry. Marginal images.**[1]

Consider a fixed plane $F$ and a plane $E$. From any point $P$ on $E$ drop a perpendicular $PQ$ on $F$. The assem-

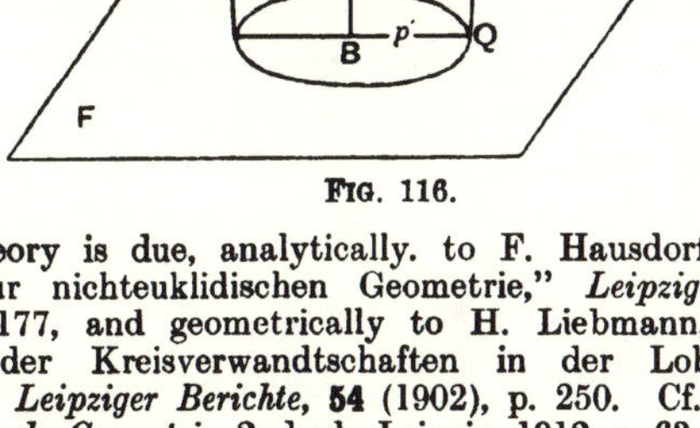


FIG. 116.

[1] This theory is due, analytically. to F. Hausdorff, "Analytische Beiträge zur nichteuklidischen Geometrie," *Leipziger Berichte*, **51** (1899),. p. 177, and geometrically to H. Liebmann, "Synthetische Ableitung der Kreisverwandtschaften in der Lobatschefskijschen Geometrie," *Leipziger Berichte*, **54** (1902), p. 250. Cf. also Liebmann, *Nichteuklidische Geometrie*, 2nd ed., Leipzig, 1912, p. 63.

blage of all points $Q$ lie within a curve called the *marginal image* of the plane $E$ on the plane $F$.

(1) Let $E$ and $F$ be non-intersecting, and have a common perpendicular $AB$ (Fig. 116). Through $A$ in the plane $E$ draw any line $AP$, and in the plane $PAB$, which cuts $F$ in $BQ$, draw $QP \perp BQ$ and $\parallel AP$. Then $Q$ lies on the marginal image of $E$.

If $AB = p$ and $BQ = p'$, then $\sinh p \sinh p' = 1$. Hence $p'$ is constant, and the marginal image is a circle with centre $B$ and radius $p'$ given by $\sinh p \sinh p' = 1$.

(2) Let $E$ cut $F$ at an angle $\alpha$ in the line $MN$ (Fig. 117). Draw a plane $\perp$ $MN$, cutting $E$ in $MP$ and $F$ in $MQ$.

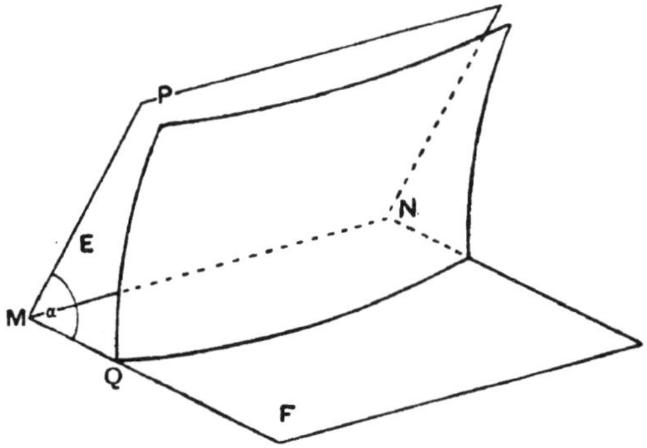


FIG. 117.

Draw $QP \perp MQ$ and $\parallel MP$. Then $\alpha = \Pi(MQ)$. Hence $MQ$ is constant, and the locus of $Q$ is an equidistant-curve with axis $MN$ and distance $a$ such that $\Pi(a) = \alpha$.

(3) Let $E$ be parallel to $F$. Then the line $MN$ goes to infinity, and the equidistant-curve becomes a horocycle.

Hence there is a (1, 1) correspondence between the circles in a plane and the planes in space.

**15.** *When the planes $E$, $E'$ intersect in a straight line $l$, their marginal images intersect in two points which form the marginal image of the line $l$.*

Let the planes $E$, $E'$ cut the absolute in conics $C$, $C'$, and let $O$ be the absolute pole of the plane $F$. Then the marginal images $I$, $I'$ of $E$, $E'$ are the projections of $C$, $C'$ on the plane $F$ with centre of projection $O$. The conics $C$, $C'$ intersect only in two points $P$, $Q$, the points of intersection of $l$ with the absolute. The cones $OC$, $OC'$ cut the absolute each in a second conic $C_1$, $C_1'$.

Now $C, C'$ cut in $P, Q$; $C_1, C_1'$ cut in $P_1, Q_1$;
$C, C_1'$ ,, $R, S$; $C', C_1$ ,, $R_1, S_1$,

and the points $P$, $P_1$ give the same projection on $F$, and so also do the other pairs. Hence the marginal images cut in four points, two of which form the marginal image of the line of intersection $l$.

**16.** *The angle between two planes is equal to the angle of intersection of their marginal images.*

Let $E_1$, $E_2$ be two planes cutting in $TT'$. Let $P_1Q_1$ be

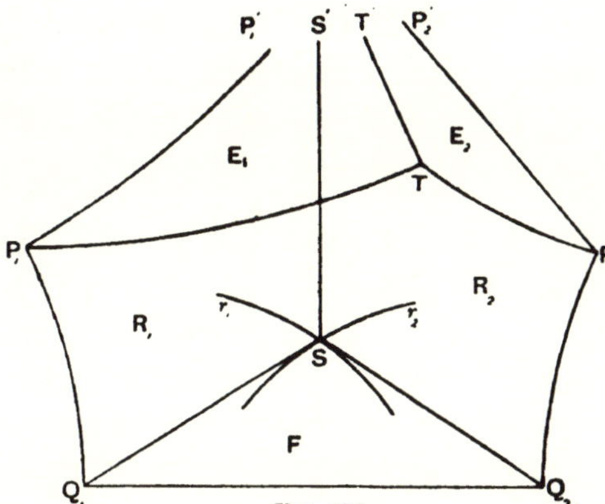


FIG. 118.

the common perpendicular of $E_1$ and $F$, $P_2Q_2$ that of $E_2$ and $F$. The plane $P_1Q_1Q_2P_2 \perp F$ cuts $TT'$ in $T$. Draw $P_1P_1'$, $P_2P_2'$ in the planes $E_1$, $E_2 \parallel TT'$, and $SS' \perp F$ and $\parallel TT'$. The planes $P_1Q_1S \equiv R_1$ and $P_2Q_2S \equiv R_2$ are $\perp F$. The marginal images $r_1$, $r_2$ of $E_1$, $E_2$ are circles with centres $Q_1$, $Q_2$ and intersecting in $S$.

We have then four planes $E_1$, $E_2$, $R_2$, $R_1$ whose lines of intersection $TT'$, $P_2P_2'$, $SS'$, $P_1P_1'$ are parallel.

Therefore $\quad \angle(E_1E_2)+(E_2R_2)+(R_2R_1)+(R_1E_1)=2\pi.$

But $\quad (R_1E_1)=\frac{\pi}{2}=(E_2R_2) \quad$ and $\quad (R_1R_2)=\pi-(r_1r_2).$

Therefore $\qquad (E_1E_2)=(r_1r_2).$

## 17. Systems of circles.

A pencil of planes through a line $l$ is represented on $F$ by a system of circles through two fixed points, the marginal image of $l$. The planes perpendicular to $l$ form a pencil of planes with ideal axis $l'$, the absolute polar of $l$. These are represented on $F$ by a system of circles through two imaginary fixed points, the marginal image of $l'$, and every circle of the first system cuts orthogonally every circle of the second system. These form therefore conjugate systems of coaxal circles.

A bundle of planes through a point $P$ is represented by a system of circles any two of which intersect in a pair of points which are the marginal image of a line through $P$. If $O$ is the foot of the perpendicular from $P$ on $F$, the two points of each pair lie on a line through $O$ and are equidistant from $O$. $O$ is the radical centre of the system, and all the circles cut orthogonally the circle which is the marginal image of the polar plane of $P$.

## 18. Types of pencils of circles.

(1) Let the axis $l$ of the pencil of planes be a non-intersector of the plane $F$, and let $PO$ be the common perpendicular of $l$ and $F$. Let $A$, $B$ be the marginal image of $l$; $O$ is the middle point of $AB$. Then the marginal images of the planes through $l$ are circles through the real points $A, B$. One of these is the line $AB$; then, as the plane $E$ is tilted, we get branches of equidistant-curves, then a horocycle, and lastly circles, ending with the circle on $AB$ as diameter, the circle of least diameter.

(1′) In the conjugate system the planes $E$ are all perpendicular to a fixed line $l$, and the axis $l'$ is ideal. The marginal images are first a straight line through $O \perp AB$, then a series of equidistant-curves with axes $\perp AB$ and increasing distances, then a horocycle, and lastly a series of circles with diminishing radii tending to the limiting point $A$; and a similar series tending to the other limiting point $B$.

(2) Let the axis $l$ cut $F$ in $O$. The marginal images are first the straight line $AB$, then a series of equidistant-curves with axes through $O$, one branch passing through each of the points $A$, $B$, ending with the equidistant-curve of greatest distance whose axis $\perp AB$.

(2*a*) If $l \perp F$, $A$ and $B$ coincide, and the marginal images are concurrent straight lines through $O$.

(2*b*) If $l$ lies in $F$, $A$ and $B$ are at infinity in opposite senses. The marginal images are equidistant-curves with common axis $AB$.

(2′) The conjugate system to (2) is similar to (1′), but instead of starting with a straight line we have first an equidistant-curve with a minimum distance

(2′*a*) When $l \perp F$ the limiting points coincide and the marginal images become concentric circles.

(2′*b*) When $l$ lies in $F$ the limiting points are at infinity in opposite senses, and the marginal images are straight lines $\perp AB$.

(3) Let $l \parallel F$, then one of the points, $B$ say, is at infinity. The marginal images are equidistant-curves through $A$ with axes parallel to $AB$, one being the straight line $AB$ and one the horocycle $\perp AB$.

(3′) In the conjugate system the limiting point $B$ is at infinity. We have first a series of equidistant-curves with increasing distances, then a horocycle, and lastly a series of circles with diminishing radii ending with the limiting point $A$.

(4) Let $l$ be at infinity with $P$ as point at infinity, and suppose $P$ is not on $F$. The planes $E$ are all parallel. The marginal image of $l$ is a point $A$, *i.e.* $A$, $B$ coincide. $A$ is the orthogonal projection of $P$ on $F$. There is a real plane through $l \perp F$ cutting $F$ in a line $t$. We have then, as the marginal images of the planes $E$, first the straight line $t$, then a series of equidistant curves, then a horocycle, and finally a series of circles, all touching $t$ at $A$, which is both a limiting point and a common point.

(4′) The conjugate system is of exactly the same form, since the absolute polar of the line $l$ at infinity touching the absolute at $P$ is also a line touching the absolute at $P$. The marginal images all touch a line $\perp t$, $A$ being the point of contact.

(5) In (4) let $P$ be on $F$ so that $A$ coincides with $P$ at infinity. The parallel planes $E$ make a constant angle $\alpha$ with $F$. We have then, as marginal images, a series of equidistant-curves with axes parallel to the direction through $A$, and constant distance $a$, such that $\Pi(a)=\alpha$.

(5$a$) If the conjugate axis $l'$ lies in $F$, $\alpha=\frac{1}{2}\pi$, and the equidistant-curves reduce to a system of parallel straight lines.

(5$b$) If $l$ lies in $F$, $\alpha=0$, and the marginal images are a system of concentric horocycles.

(5′) The conjugate system to (5) is a system of exactly the same form with the angle $\alpha'=\dfrac{\pi}{2}-\alpha$.

(5′$a$) is the same as (5$b$) and (5′$b$) the same as (5′$a$).

*Note.*—In (2$a$) we appear to have a pencil of circles with coincident common points, but we must consider this actually as a pencil with one real common point, and an ideal common point which is the inverse of $O$ with respect to the absolute. Similarly in (2) we should regard the two branches of the equidistant-curves separately; and regard the whole system as consisting of two pencils, each with one actual and one ideal common point. (Cf. Ex. VIII. 19, 20.) This is rendered clearer if, in finding the marginal images, we confine our attention to the parts of the planes and lines which lie on one side of the plane $F$.

## EXAMPLES VII.

**1.** Prove that if the common tangents to two circles are all real, the distances between the points of contact are equal in pairs, and that all four distances will be equal only if the axes of the circles are orthogonal.

**2.** Prove that the second radical axis of two circles which pass through $A$, $B$ passes through the middle point of one of the segments $AB$, $BA$.

3. Prove that $x+\lambda y-kz=0$ represents, for parameter $\lambda$, a pencil of lines parallel to the positive axis of $x$.

**4.** In elliptic geometry, prove that the circles $a_1x+b_1y+c_1z\pm d_1=0$ and $a_2x+\text{etc.}=0$ will cut orthogonally if $k^2(a_1a_2+b_1b_2)+c_1c_2\pm d_1d_2=0$.

**5.** If a bundle of circles contains a pencil of lines parallel to the positive axis of $x$, show that the equation of the bundle can be written in the form $(x+pk)+\lambda y+\mu(z+p)=0$.

**6.** If a bundle of circles contains a pencil of lines perpendicular to the axis of $x$, show that the equation of the bundle can be written $x+\lambda(y+b)+\mu z=0$.

**7** Prove that the orthogonal circle of the bundle of circles $x+\lambda y+\mu(z+c)=0$ is $cz=1$.

**8.** Prove that every circle of the system $(x-pk)+\lambda y+\mu(z-p)=0$ cuts orthogonally the horocycle $p(x-kz)=k$.

**9.** If the orthogonal circle of the bundle of circles

$$x+\lambda y+\mu(z+c)=0$$

is imaginary, prove that every circle of the system passes through the ends of a diameter of the fixed circle $z+c=0$.

**10.** Prove that the locus of the centres of point-circles of the bundle $x+\lambda(y+b)+\mu z=0$ is the equidistant-curve $by=\pm k^2$.

**11.** If $\sinh p \sinh p'=1$, prove that

$$\Pi(p)+\Pi(p')=\frac{\pi}{2}, \quad \text{and} \quad p'=\log\coth\frac{p}{2}.$$

**12.** Given a circle, equidistant-curve or horocycle in a plane $F$, show how to construct the plane $E$ of which it is the marginal image on the plane $F$.

# CHAPTER VIII.

## INVERSION AND ALLIED TRANSFORMATIONS.[1]

**1.** In euclidean geometry, Inversion, or the transformation by reciprocal radii, is a transformation which changes any point $P$ into a point $P'$, and $P'$ into $P$; the line $PP'$ passes through a fixed point $O$, the centre of inversion, and the segments $OP$, $OP'$ are connected by the relation $OP \,.\, OP' = \text{constant}$. This transformation has the properties that it changes circles into circles and transforms angles unaltered in magnitude. It is a special case of a *conformal transformation* which preserves angles, and of the more special type of conformal transformation, the *circular transformation* which changes circles into circles.

We shall consider in this chapter the circular transformations in the non-euclidean plane, and first we shall prove the following theorem.

**2.** *Any point-transformation which changes circles into circles is conformal.*

Two circles which intersect at equal angles at $A$, $B$ are transformed into two circles which intersect at equal angles at $A'$, $B'$, *i.e.* certain pairs of equal angles are transformed into pairs of equal angles. We shall show that this holds for all pairs of equal angles.

[1] See the references to Hausdorff and Liebmann in chap. vii. § 14.

Let the lines $a_1$, $b_1$ through $S_1$ and $a_2$, $b_2$ through $S_2$ make equal angles in the same sense. Let $a_1$, $a_2$ meet in $O$,

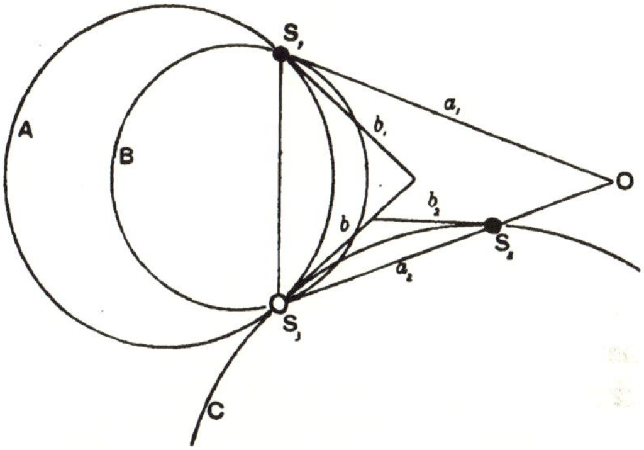


FIG. 119.

and let $S_1S_3$ make equal angles with $a_1$ and $a_2$. Draw $b$ through $S_3$ making $\widehat{a_1b_1}=\widehat{ba_2}$, then $\widehat{ba_2}=\widehat{a_2b_2}$.

Two circles $A$, $B$ can be drawn with their centres on the bisector of the angle at $O$, passing through $S_1$ and $S_3$ and having $a_1$, $a_2$ and $b_1$, $b$ as tangents ; and similarly a circle $C$ can be drawn with its centre on the perpendicular bisector of $S_3S_2$ and having $b$, $b_2$ as tangents.

The equal angles $\widehat{a_1b_1}$ and $\widehat{ba_2}$ are transformed into equal angles, and the equal angles $\widehat{ba_2}$ and $\widehat{a_2b_2}$ are also transformed into equal angles, *i.e.* a pair of equal angles in any position are transformed into a pair of equal angles.

Hence two adjacent right angles are transformed into two adjacent right angles, half a right angle is transformed into half a right angle, and so on. Hence angles are unchanged.

**3**. Consider two planes $F_1$, $F_2$. All the planes in space are represented on each of these by circles, and we have a correspondence between the circles of $F_1$ and the circles of $F_2$ through the medium of the planes of space. Then, if $F_2$ is made to coincide with $F_1$, we have established a correspondence between the circles of $F_1$ itself, *i.e.* we have effected a transformation of $F_1$, changing circles into circles.

Instead of supposing the plane $F_2$ to move, we may suppose $F$ to be a fixed plane and let the whole of space move rigidly, with the exception of $F$. To a circle $C$ corresponds a plane $E$. This plane is moved to $E'$ and gives another circle $C'$. To a pencil of circles corresponds an axial pencil of planes, and this gives again a pencil of circles. To a bundle of circles with common radical centre $O$ corresponds a bundle of planes through a point $P$; $P$ is moved to $P'$, and we get another bundle of circles with common radical centre $O'$. Hence this effects a transformation of the plane $F$, changing a point into a point, and a circle into a circle. It does not change a straight line into a straight line, but in general into a circle.

The motion of space which has just been considered is a kind of *congruent transformation*, *i.e.* it does not alter distances or angles. But a congruent transformation considered more generally may reverse the order of objects, changing, for example, a right-hand glove into a left-hand glove. Such a transformation is produced by a *reflexion* in a plane. A motion is equivalent to two reflexions.

We may extend the above result, therefore, and say: *Every congruent transformation of space gives rise to a circular transformation of a plane.*

**4.** Conversely: *Every point-transformation of the plane which changes circles into circles can be represented by a congruent transformation of space.*

To a circle $C$ corresponds a plane $E$, and to the corresponding circle $C'$ corresponds a plane $E'$. Hence a plane is transformed into a plane, and the angle between two planes is equal to the angle between the corresponding planes. Further, a pencil of circles is transformed into a pencil of circles (since the transformation is a point-transformation); hence a straight line, the axis of a pencil of planes, is transformed into a straight line. Also a bundle of circles is transformed into a similar system; hence a point, the vertex of a bundle of planes, is transformed into a point. The transformation of space therefore changes points, lines and planes into points, lines and planes, and leaves angles unaltered, *i.e.* it is a congruent transformation.[1]

**5.** The general circular transformation which we have been considering is more general than inversion, for inversion leaves unaltered a point $O$, the centre of inversion, and also all straight lines through $O$.

In general a system of lines through a point is transformed into a pencil of circles. In a pencil of circles through two points $A$, $B$ there is always one straight line, the straight line $AB$; and if a pencil of circles contains two straight lines it must consist entirely of straight lines; for the planes corresponding to the two lines are both perpendicular to $F$, and any plane through their line of intersection is also perpendicular to $F$.

Now a pencil of lines through a point $A$ is transformed into a pencil of circles through $A'$, $B'$. Hence one line of the pencil is transformed into the straight line $A'B'$. Hence

[1] In euclidean space these conditions would specify only a *similar* transformation. In non-euclidean geometry, when the angles of a triangle are given, its sides are also determined.

through any point $A$ there is one straight line $g$ which is transformed into a straight line $g'$. Let $h$ be another line which is transformed into a straight line $h'$. Then the pencil $(gh)$, which consists entirely of straight lines, is transformed into a pencil $(g'h')$, which consists entirely of straight lines. If $g$, $h$ intersect in $O$, then $g'$, $h'$ intersect in $O'$, and

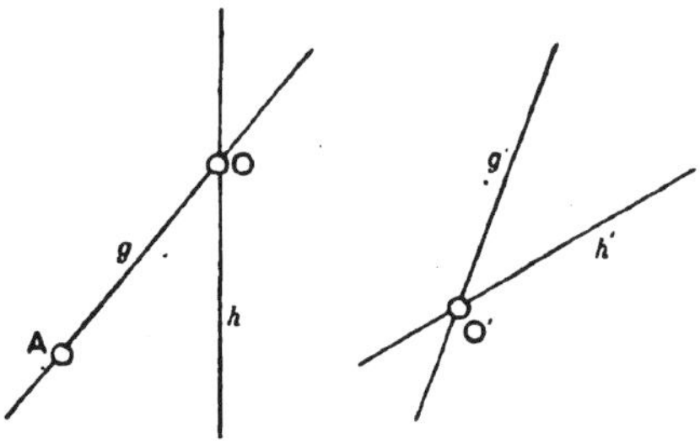


FIG. 120.

the corresponding angles at $O$ and $O'$ are equal. Let $O$ be moved into coincidence with $O'$ and $g$ with $g'$. Then either $h$ and $h'$ coincide or can be brought into coincidence by flapping the whole plane over about $g'$, *i.e.* by a reflexion in $g'$. Then all the other lines of the first pencil will coincide with their correspondents, since angles are unaltered. Hence, *the general circular transformation is compounded of a congruent transformation of the plane and a circular transformation which leaves unaltered all the straight lines through a fixed point.*

**6.** Of this simpler form of circular transformation, which keeps one point fixed, there are three types, according as the fixed point is real, ideal or at infinity. These are called the *hyperbolic*, *elliptic* and *parabolic* types.

And further, there are two forms of each, according as

the corresponding congruent transformation of space is a reflexion or a motion.

In the first case points are connected in pairs, since the relation between a point and its image is symmetrical. If $P$ is transformed into $P'$, then by the same transformation $P'$ is transformed into $P$. A repetition of the transformation will reproduce the *status quo*. The transformation is therefore periodic with period 2, or, as it is called, involutory. This form of transformation is called an *inversion*.

In the other case, by repeated transformation the transformed points on a fixed line always go in the same direction. This form is called a *radiation*.

7. We shall now determine the metrical relations which define inversion.

(1) *Hyperbolic Inversion*, with real centre $O$. Draw a line $OD=d$ perpendicular to the plane $F$, and through $D$

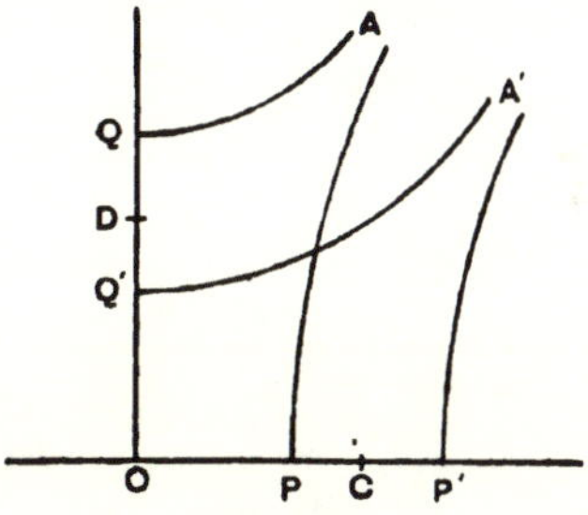


FIG. 121.

draw a plane $K \perp OD$. We shall obtain a hyperbolic inversion by a reflexion of space in the plane $K$. Take any point $P$ in $F$ and draw $PA \perp F$, and $QA \perp OD$ and $\parallel PA$. Let $Q'$ be the reflexion of $Q$, so that $DQ'=QD$, and in the plane $POQ$ draw $Q'A' \perp OD$, and $P'A' \perp F$

and $\parallel Q'A'$. Then $P'$ is the point which corresponds to $P$. Construct the point $C$ which corresponds to $D$. Let $OC=c$, $QD=DQ'=y$, $OP=x$, $OP'=x'$.

Then

$$\sinh c \sinh d = 1 = \sinh x \sinh(d+y) = \sinh x' \sinh(d-y);$$

therefore $\sinh x = \operatorname{cosech}(d+y)$ and $\cosh x = \coth(d+y)$.

$$\tanh \tfrac{1}{2}x = \coth x - \operatorname{cosech} x$$
$$= \cosh(d+y) - \sinh(d+y) = e^{-(d+y)}.$$

Similarly $\tanh \tfrac{1}{2}x' = e^{-(d-y)}$ and $\tanh \tfrac{1}{2}c = e^{-d}$.

Hence $\tanh \tfrac{1}{2}x \tanh \tfrac{1}{2}x' = e^{-2d} = \tanh^2 \tfrac{1}{2}c.$

This is the formula for inversion in a circle of radius $c$.

(2) *Elliptic Inversion*, with ideal centre O. The fixed lines are all perpendicular to a fixed line $l$. Draw a plane

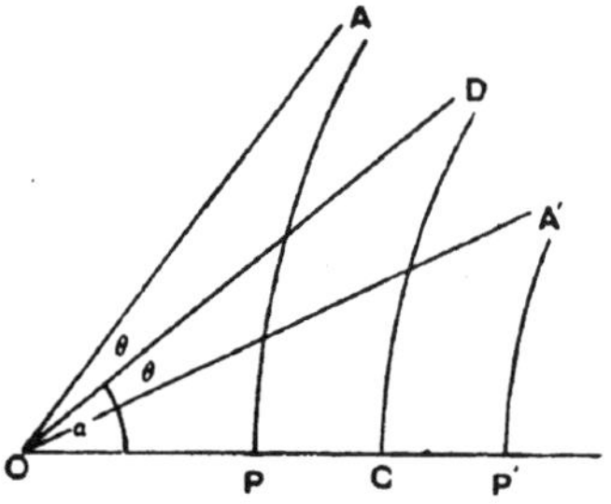


FIG. 122.

$K$ through $l$ making an angle $a$ with $F$, and take this as the plane of reflexion. Then in Fig. 122, where

$$AOD = DOA' = \theta,$$

$PA \perp OP$ and $\parallel OA$, $P'A' \perp OP'$ and $\parallel OA'$, $OP=x$, $OP'=x'$, $OC=c$, we have

$$\Pi(x) = a+\theta, \quad \Pi(x') = a-\theta.$$

Therefore $\Pi(x) + \Pi(x') = 2a = 2\Pi(c)$.

This is the formula for inversion in an equidistant-curve of distance $c$.

If $a = \frac{\pi}{2}$, this gives $x' = -x$, a reflexion in a straight line.

If $a = \frac{\pi}{4}$, we have $\Pi(x) + \Pi(x') = \frac{\pi}{2}$, or $\sinh x \sinh x' = 1$, a form of transformation which was frequently used by Lobachevsky in establishing the trigonometrical formulae.

(3) *Parabolic Inversion*, with centre $\Omega$ at infinity. The corresponding congruent transformation of space consists of a reflexion in a plane $K \parallel F$.

In Fig. 123 $XD$ is the trace of the fixed plane $K$, $C$ the marginal image of $D$; $UA$ is the trace of a plane $\parallel F$, $U'A'$

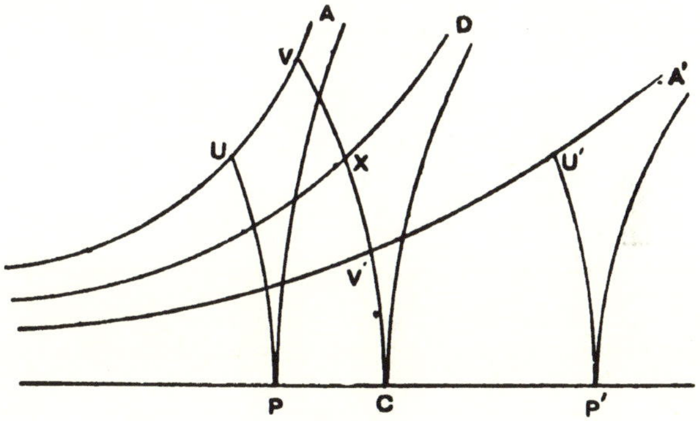


FIG. 123.

the trace of the reflexion of $UA$ in $K$, and $P$, $P'$ the marginal images of $A$ and $A'$.

Draw the horocyclic arcs $PU$, $CV$, $P'U'$. Let $CP = x$, $CP' = x'$, $x$ being positive and $x'$ negative.

Then $PU = CX = P'U' = k = 1$, since each is the arc of a

horocycle having the tangent at one end parallel to the radius at the other end.

$$XV = XV', \quad CV = PU \, . \, e^x, \quad CV' = P'U' \, . \, e^{x'};$$

also

$$CV + CV' = 2CX.$$

Therefore

$$e^x + e^{x'} = 2.$$

This is the formula for inversion in a horocycle.

8. There is one property in which non-euclidean inversion differs from euclidean. In euclidean inversion the inverse $P'$ of a point $P$ with respect to a circle of radius $OA$ is the

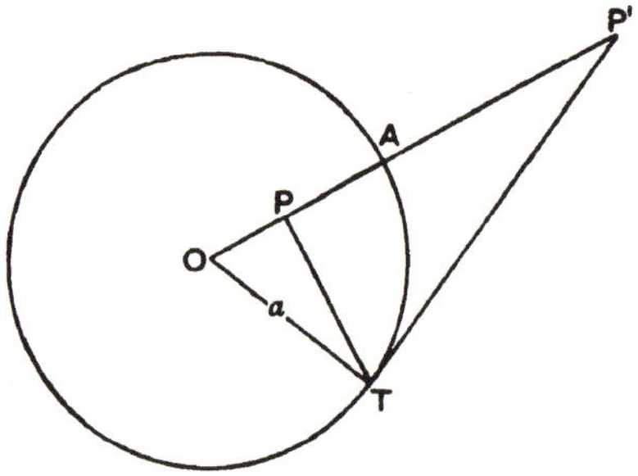


FIG. 124.

point of intersection of the radius $OP$ with the polar of $P$. This does not hold in non-euclidean geometry.

If $P'T$ is a tangent to the circle, and $OP = r$, $OP' = r'$, we have

$$\cos TOP' = \coth OP' \tanh a = \tanh OP \coth a.$$

Hence

$$\tanh r \tanh r' = \tanh^2 a,$$

whereas the distances of the inverse points are connected by the relation

$$\tanh \tfrac{1}{2}r \tanh \tfrac{1}{2}r' = \tanh^2 \tfrac{1}{2}a.$$

In euclidean geometry these both reduce to the same, $rr' = a^2$.

The transformation which is called in euclidean geometry "quadric inversion," and which is obtained by the above construction with the circle replaced by any conic, is, therefore, in non-euclidean geometry, not a generalization of inversion.

**9. Congruent transformations. Transformation of coordinates.**

The equations which determine a congruent transformation of the plane are given at once by the equations for transformation of coordinates.

Let the rectangular axes $Ox$, $Oy$ be moved into the position $O'x'$, $O'y'$, still remaining rectangular. Let the

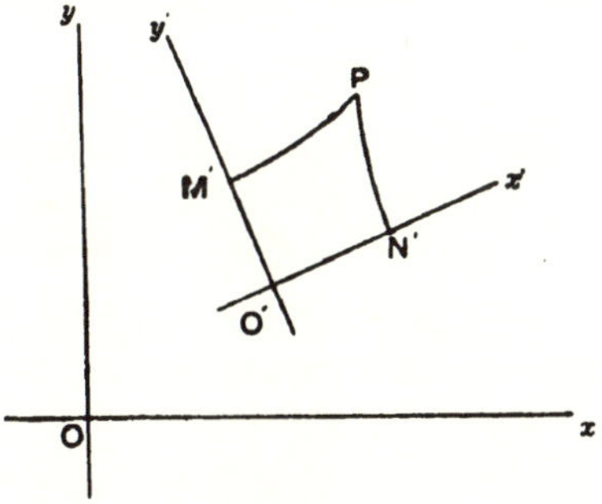


FIG. 125.

coordinates of $O'$ be $(a_0, b_0, c_0)$, and the equations of $O'x'$, $O'y'$,

$$a_2x + b_2y + c_2z = 0,$$
$$a_1x + b_1y + c_1z = 0.$$

Then, the geometry being hyperbolic,

$$\left.\begin{aligned} a_1^2 + b_1^2 - c_1^2/k^2 &= 1, \\ a_2^2 + b_2^2 - c_2^2/k^2 &= 1, \\ a_0^2 + b_0^2 - k^2c_0^2 &= -k^2. \end{aligned}\right\} \qquad (1)$$

Also, since the new axes are rectangular and pass through $O'$,

$$\left.\begin{aligned} a_1a_2+b_1b_2-c_1c_2/k^2&=0,\\ a_1a_0+b_1b_0+c_1c_0\quad&=0,\\ a_2a_0+b_2b_0+c_2c_0\quad&=0;\end{aligned}\right\}\quad\ldots\ldots\ldots\ldots(2)$$

whence

$$a_0:b_0:c_0:1=b_1c_2-b_2c_1:c_1a_2-c_2a_1:a_1b_2-a_2b_1:R.$$

To determine the factor $R$, we have

$$\begin{aligned}-k^2R^2&=(b_1c_2-b_2c_1)^2+(c_1a_2-c_2a_1)^2-k^2(a_1b_2-a_2b_1)^2\\ &=-k^2(a_1^2+b_1^2-c_1^2/k^2)(a_2^2+b_2^2-c_2^2/k^2)\\ &\qquad\qquad+k^2(a_1a_2+b_1b_2-c_1c_2/k^2)^2\\ &=-k^2.\end{aligned}$$

Therefore $$R=\pm 1.$$

There are two cases in the transformation, according as the new axes follow the same order as the old, or are reversed. The second case is obtained from the first by interchanging $x$ and $y$. If the axes are supposed to be fixed while the whole plane moves, we call these two cases respectively a *motion* and a *reflexion* of the plane. A reflexion may be produced by flapping the plane over about a line in it.

If the axes are unchanged we have $a_2=0$, $b_2=1$, $a_1=1$, $b_1=0$, $c_0=1$, and therefore $a_1b_2-a_2b_1=1$. If the axes are interchanged $a_2=1$, $b_2=0$, $a_1=0$, $b_1=1$, $c_0=1$, and therefore $a_1b_2-a_2b_1=-1$. Hence for a motion $R=+1$, for a reflexion $R=-1$.

We have then

$$\begin{aligned} &b_1c_2-b_2c_1=Ra_0, && c_1a_2-c_2a_1=Rb_0, && a_1b_2-a_2b_1=Rc_0,\\ &b_1c_0+b_0c_1/k^2=-Ra_2, && a_1c_0+a_0c_1/k^2=Rb_2, && a_1b_0-a_0b_1=-Rc_2,\\ &b_2c_0+b_0c_2/k^2=Ra_1, && a_2c_0+a_0c_2/k^2=-Rb_1, && a_2b_0-a_0b_2=Rc_1.\end{aligned}$$

**10.** Let $P$ be any point, whose coordinates referred to the old and the new axes are $(x, y, z)$ and $(x', y', z')$. Then, expressing the distances of $P$ from $O'x'$, $O'y'$ and $O'$, we have $x' = k \sinh M'P/k$, $y' = k \sinh N'P/k$, $z' = \cosh O'P/k$; hence

$$\left.\begin{aligned} x' &= a_1 x + b_1 y + c_1 z, \\ y' &= a_2 x + b_2 y + c_2 z, \\ -k^2 z' &= a_0 x + b_0 y - k^2 c_0 z, \end{aligned}\right\} \quad \ldots\ldots\ldots\ldots (\text{A})$$

where the nine coefficients are connected by the six relations (1) and (2).

Further, since $x'^2 + y'^2 - k^2 z'^2 = -k^2$, we have

$$\begin{aligned} a_1^2 + a_2^2 - a_0^2/k^2 &= 1, & b_1 c_1 + b_2 c_2 + b_0 c_0 &= 0, \\ b_1^2 + b_2^2 - b_0^2/k^2 &= 1, & c_1 a_1 + c_2 a_2 + c_0 a_0 &= 0, \\ c_1^2 + c_2^2 - k^2 c_0^2 &= -k^2, & a_1 b_1 + a_2 b_2 - a_0 b_0/k^2 &= 0. \end{aligned}$$

Multiplying the equations (A) respectively by $a_1$, $a_2$, $-a_0/k^2$ and adding, we get

Similarly
$$\left.\begin{aligned} x &= a_1 x' + a_2 y' + a_0 z'. \\ y &= b_1 x' + b_2 y' + b_0 z', \\ -k^2 z &= c_1 x' + c_2 y' - k^2 c_0 z', \end{aligned}\right\} \quad \ldots\ldots\ldots\ldots (\text{A}')$$

from which we see that the coordinates of $Oy'$, $Ox'$, $O$ referred to the new axes are

$$(a_1, a_2, a_0), \quad (b_1, b_2, b_0), \quad (c_1, c_2, c_0).$$

(A′) is the inverse transformation to (A). Both can be represented by the scheme

| | $x$ | $y$ | $ikz$ |
|---|---|---|---|
| $x'$ | $a_1$ | $b_1$ | $c_1/ik$ |
| $y'$ | $a_2$ | $b_2$ | $c_2/ik$ |
| $ikz'$ | $a_0/ik$ | $b_0/ik$ | $c_0$ |

which may be read either horizontally or vertically.

The determinant of the substitution $= +1$ for a motion, $-1$ for a reflexion.

**11.** The transformation admits of a very simple representation.

Since $x^2+y^2-k^2z^2=-k^2$, we can write

$$(x+iy)(x-iy)=k^2(z-1)(z+1).$$

Let $$\lambda=\frac{x+iy}{k(z+1)}=\frac{k(z-1)}{x-iy},\quad \bar{\lambda}=\frac{x-iy}{k(z+1)}=\frac{k(z-1)}{x+iy}.$$

Then $$\lambda\bar{\lambda}=\frac{z-1}{z+1},\quad 1-\lambda\bar{\lambda}=\frac{2}{z+1}$$

and $$x=k\frac{\bar{\lambda}+\lambda}{1-\lambda\bar{\lambda}},\quad y=ik\frac{\bar{\lambda}-\lambda}{1-\lambda\bar{\lambda}},\quad z=\frac{1+\lambda\bar{\lambda}}{1-\lambda\bar{\lambda}}.$$

The coordinates of a point are then expressed in terms of the complex parameter $\lambda$, just as in Argand's diagram.

**12.** Let $\lambda'$ be the parameter of $P$ referred to the new axes. We have to express $\lambda'$ in terms of $\lambda$.

$$x'+iy'=(a_1+ia_2)x+(b_1+ib_2)y+(c_1+ic_2)z,$$
$$k(z'+1)=(k^2-a_0x-b_0y+k^2c_0z)/k.$$

Multiplying these by $1-\lambda\bar{\lambda}$, we get

$$N=k(a_1+ia_2)(\bar{\lambda}+\lambda)+ik(b_1+ib_2)(\bar{\lambda}-\lambda)$$
$$+(c_1+ic_2)(1+\lambda\bar{\lambda})$$
$$=k\lambda[(a_1+ia_2)-i(b_1+ib_2)]$$
$$+k\bar{\lambda}[(a_1+ia_2)+i(b_1+ib_2)]+(1+\lambda\bar{\lambda})(c_1+ic_2).$$
$$D=k(1-\lambda\bar{\lambda})-a_0(\bar{\lambda}+\lambda)-ib_0(\bar{\lambda}-\lambda)+kc_0(1+\lambda\bar{\lambda}).$$

Now $$(a_0-ib_0)[(a_1+ia_2)+i(b_1+ib_2)]$$
$$=a_0a_1+b_0b_1-a_0b_2+b_0a_2+i(a_0a_2+b_0b_2+a_0b_1-b_0a_1)$$
$$=-c_0c_1+Rc_1+i(-c_0c_2+Rc_2)=-(c_0-R)(c_1+ic_2),$$

and $$(a_0+ib_0)[(a_1+ia_2)-i(b_1+ib_2)]=-(c_0+R)(c_1+ic_2).$$

Let $\lambda_0$ be the parameter of $O'$, so that

$$\lambda_0 = \frac{k(c_0-1)}{a_0-ib_0}, \quad \frac{1}{\lambda_0} = \frac{k(c_0+1)}{a_0+ib_0}.$$

Then, for a motion, $R=+1$, and

$$N\lambda_0 = (c_1+ic_2)(\lambda_0-\lambda)(1-\lambda_0\bar{\lambda}),$$
$$D = k(c_0+1)(1-\lambda\bar{\lambda}_0)(1-\lambda_0\bar{\lambda}).$$

Now $$c_1^2+c_2^2 = k^2(c_0^2-1);$$

therefore $c_1+ic_2 = -k\sqrt{c_0^2-1}e^{i\phi}$, where $\phi = \pi+\tan^{-1}\frac{c_2}{c_1}$,

and $$a_0^2+b_0^2 = k^2(c_0^2-1);$$

therefore $\lambda_0(c_0+1) = \sqrt{c_0^2-1}e^{i\psi}$, where $\psi = \tan^{-1}\frac{b_0}{a_0}$.

Therefore $N = k(c_0+1)(\lambda-\lambda_0)(1-\lambda_0\bar{\lambda})e^{i(\phi-\psi)}$.

Hence $$\lambda' = \frac{N}{D} = \frac{\lambda-\lambda_0}{1-\lambda\bar{\lambda}_0}e^{i(\phi-\psi)}.$$

Let $e^{i(\phi-\psi)} = \alpha/\bar{\alpha}$, and $\lambda_0\alpha = -\beta$, $\bar{\lambda}_0\bar{\alpha} = -\bar{\beta}$; then

$$\lambda' = \frac{\alpha\lambda+\beta}{\bar{\beta}\lambda+\bar{\alpha}},$$

i.e. *the general transformation of coordinates, or the general motion in the plane, can be represented by a type of homographic transformation of a complex parameter.*

**13.** If $S$ denotes the operation which changes $\lambda$ into $\lambda'$ by the above equation, then the product of two such operations $S_1$ and $S_2$ leads to

$$\lambda'' = \frac{\alpha_2\lambda'+\beta_2}{\bar{\beta}_2\lambda'+\bar{\alpha}_2}$$
$$= \frac{\alpha_2(\alpha_1\lambda+\beta_1)+\beta_2(\bar{\beta}_1\lambda+\bar{\alpha}_1)}{\bar{\beta}_2(\alpha_1\lambda+\beta_1)+\bar{\alpha}_2(\bar{\beta}_1\lambda+\bar{\alpha}_1)} = \frac{\alpha\lambda+\beta}{\bar{\beta}\lambda+\bar{\alpha}},$$

where $\alpha = \alpha_1\alpha_2+\bar{\beta}_1\beta_2$, $\beta = \alpha_2\beta_1+\bar{\alpha}_1\beta_2$;

therefore $S_1S_2$ is an operation of the same form. The operations $S$ have therefore the property that the product of any two of them is again an operation $S$.

Further, it can be proved that $(S_1S_2)S_3 = S_1(S_2S_3)$, and the operations are associative. A set of operations satisfying these two conditions is called a *group*. This group is called the *group of non-euclidean motions*.

The homographic transformation which represents a motion is a particular case of the general homographic transformation

$$\lambda' = \frac{\alpha\lambda + \beta}{\gamma\lambda + \delta},$$

where $\alpha$, $\beta$, $\gamma$, $\delta$ are any complex numbers. This transformation belongs to a more general group, the *group of homographic transformations*, and the group of motions is a *sub-group* of this larger group.

**14.** In elliptic geometry a motion is represented by the transformation

$$\lambda' = \frac{\alpha\lambda - \beta}{\bar{\beta}\lambda + \bar{\alpha}}.$$

If $S$ is the product of two operations $S_1$, $S_2$, we have

$$\alpha = \alpha_1\alpha_2 - \bar{\beta}_1\beta_2, \qquad \beta = \alpha_2\beta_1 + \bar{\alpha}_1\beta_2.$$

Put $\alpha = d + \iota a$, $\beta = b - \iota c$, where $a$, $b$, $c$, $d$ are real and $\iota = \sqrt{-1}$; then we have

$$\begin{aligned} a &= \phantom{-}a_1d_2 + b_1c_2 - c_1b_2 + d_1a_2, \\ b &= -a_1c_2 + b_1d_2 + c_1a_2 + d_1b_2, \\ c &= \phantom{-}a_1b_2 - b_1a_2 + c_1d_2 + d_1c_2, \\ d &= -a_1a_2 - b_1b_2 - c_1c_2 + d_1d_2. \end{aligned}$$

Now these relations are exactly the same as those which we obtain from the equation

$$ai + bj + ck + d = (a_1i + b_1j + c_1k + d_1)(a_2i + b_2j + c_2k + d_2),$$

where

$$i^2 = j^2 = k^2 = -1,$$

$$jk = i = -kj, \qquad ki = j = -ik, \qquad ij = k = -ji.$$

Here $ai+bj+ck+d$ $(=q)$ is a *quaternion*. Hence the rule for compounding operations of the group

$$\lambda' = (a\lambda - \beta)/(\bar{\beta}\lambda + \bar{a})$$

is exactly the same as that for quaternions. The meaning of this can be explained as follows. The operation $q(\ \ )q^{-1}$ performed upon a vector ( ) has the effect of a rotation about a definite line. The product of two such operations

$$q_2\{q_1(\ \ )q_1^{-1}\}q_2^{-1} = q_2q_1(\ \ )q_1^{-1}q_2^{-1} = q(\ \ )q^{-1},$$

where $q = q_2q_1$, and is therefore another operation of the same form. These operations form the group of rotations about a fixed point, or the group of motions on the sphere.

**15.** If we take polar coordinates $(r, \theta)$,

$$x + iy = k\sinh\frac{r}{k}(\cos\theta + i\sin\theta) = k\sinh\frac{r}{k}e^{i\theta},$$

$$z + 1 = \cosh\frac{r}{k} + 1 = 2\cosh^2\frac{r}{2k}.$$

Therefore $$\lambda = \tanh\frac{r}{2k}e^{i\theta},$$

$$\lambda\bar{\lambda} = \tanh^2\frac{r}{2k}, \quad \sqrt{\lambda\bar{\lambda}}e^{i\theta} = \lambda, \quad \cosh\frac{r}{k} = \frac{1+\lambda\bar{\lambda}}{1-\lambda\bar{\lambda}}, \quad \sinh\frac{r}{k} = \frac{2\sqrt{\lambda\bar{\lambda}}}{1-\lambda\bar{\lambda}}.$$

The equation of a straight line $ax + by + cz = 0$ becomes, when expressed in terms of $\lambda$,

$$ak(\bar{\lambda} + \lambda) + ibk(\bar{\lambda} - \lambda) + c(1 + \lambda\bar{\lambda}) = 0,$$

*i.e.* $$c(\lambda\bar{\lambda} + 1) + k(a + ib)\bar{\lambda} + k(a - ib)\lambda = 0,$$

which is of the form

$$\lambda\bar{\lambda} - \bar{a}\lambda - a\bar{\lambda} + 1 = 0.$$

If the line passes through the origin, $c = 0$, and the equation reduces to

$$\bar{a}\lambda + a\bar{\lambda} = 0.$$

The equation of a circle,

$$\cosh\frac{a}{k}=\cosh\frac{c}{k}\cosh\frac{r}{k}-\sinh\frac{c}{k}\sinh\frac{r}{k}\cos(\theta-\alpha),$$

with centre $(c, \alpha)$ and radius $a$, becomes

$$\lambda\bar{\lambda}\left(\cosh\frac{c}{k}+\cosh\frac{a}{k}\right)-\sinh\frac{c}{k}(\lambda e^{-i\alpha}+\bar{\lambda}e^{i\alpha})$$
$$+\cosh\frac{c}{k}-\cosh\frac{a}{k}=0.$$

In general, therefore, the equation

$$\lambda\bar{\lambda}-\bar{a}\lambda-a\bar{\lambda}+b=0$$

represents a circle (equidistant-curve or horocycle) which reduces to a straight line when $b=1$.

**16.** The general homographic transformation of $\lambda$ leaves the form of the equation of a circle unaltered, and is therefore a circular transformation. The transformation of inversion is included in this. Inversion is characterised by connecting points in pairs. The parameters $\lambda$, $\lambda'$ of a pair of inverse points must therefore be connected by a lineo-linear equation of one of the forms

$$\lambda'=\frac{\alpha\lambda+\beta}{\gamma\lambda+\delta},\quad \bar{\lambda}'=\frac{\bar{\alpha}\bar{\lambda}+\bar{\beta}}{\bar{\gamma}\bar{\lambda}+\bar{\delta}},\quad \text{..........(1)}$$

$$\lambda'=\frac{\alpha\bar{\lambda}+\beta}{\gamma\bar{\lambda}+\delta},\quad \bar{\lambda}'=\frac{\bar{\alpha}\lambda+\bar{\beta}}{\bar{\gamma}\lambda+\bar{\delta}}.\quad \text{..........(2)}$$

The first form characterises motions, the second reflexions, when $\gamma=\bar{\beta}$ and $\delta=\bar{\alpha}$. Inversion belongs to the second form and is a symmetrical transformation, *i.e.* $\lambda'$ is expressed in terms of $\bar{\lambda}$ by exactly the same equation as that which expresses $\lambda$ in terms of $\bar{\lambda}'$.

If $\gamma\neq 0$, we can take $\lambda'=\dfrac{\alpha\bar{\lambda}+\beta}{\bar{\lambda}+\delta}$.

Then $\bar{\lambda} = \dfrac{-\delta\lambda' + \beta}{\lambda' - a}$ and $\lambda = \dfrac{-\bar{\delta}\bar{\lambda}' + \bar{\beta}}{\bar{\lambda}' - \bar{a}}$.

Hence $\delta = -\bar{a}$ and $\beta = \bar{\beta}$, *i.e.* $\beta$ is a real number $= -b$. The transformation for inversion is therefore of the form

$$\lambda' = \frac{a\bar{\lambda} - b}{\bar{\lambda} - \bar{a}}. \quad \text{.................. (I)}$$

If the points $\lambda$, $\lambda'$ coincide, so that $\lambda' = \lambda$,

$$\lambda\bar{\lambda} - \bar{a}\lambda - a\bar{\lambda} + b = 0, \quad \text{.............. (C)}$$

which is the equation of the circle of inversion. If $\lambda$, $\lambda'$ are a pair of corresponding points, equation (I) gives

$$\lambda'\bar{\lambda} - \bar{a}\lambda' - a\bar{\lambda} + b = 0.$$

If $\gamma = 0$, we can take $\lambda' = a\bar{\lambda} + \beta$. Proceeding as before we find $a\bar{a} = 1$ and $\beta = -a\bar{\beta}$. The transformation then reduces to the form

$$\bar{a}\lambda' + a\bar{\lambda} = b.$$

In this case the circle of inversion is the inverse with regard to the absolute of a circle which passes through the origin.

These results should be compared with the corresponding formulae for euclidean geometry in Chapter V. §§ 31, 32.

## EXAMPLES VIII.

1. In elliptic geometry, show that the general transformation of coordinates is expressed by $\lambda' = \dfrac{(a+ib)\lambda - (c+id)}{(c-id)\lambda + (a-ib)}$, where $a$, $b$, $c$, $d$ are real.

2. Prove that the general homographic transformation $\lambda' = \dfrac{a\lambda + \beta}{\gamma\lambda + \delta}$ changes circles into circles.

3. Show that the transformations $\lambda' = \dfrac{a\lambda + \beta}{\gamma\lambda + \delta}$ form a group.

**4.** Show that the general reflexion of the plane in hyperbolic geometry is represented by $\lambda' = \dfrac{a\bar{\lambda}+\beta}{\bar{\beta}\bar{\lambda}+\bar{a}}$.

**5.** Show that the reflexions of the plane do not form a group, but that the product of two reflexions is a motion.

**6.** Show that the operations of the group $\lambda' = \dfrac{a\lambda+\beta}{\bar{\beta}\lambda+\bar{a}}$ leave unaltered the equation $\lambda\bar{\lambda}=1$

**7.** Show that the equation $y=0$ is unaltered by the operations of the group $\lambda' = \dfrac{a\lambda+b}{c\lambda+d}$, where $a, b, c, d$ are real.

**8.** Show that the equation $x=0$ is unaltered by the operations of the group $\lambda' = \dfrac{a\lambda+ib}{ic\lambda+d}$, where $a, b, c, d$ are real.

**9.** If the points $\lambda_1$, $\lambda_2$, $\lambda_3$ are collinear, prove that

$$\begin{vmatrix} 1+\lambda_1\bar{\lambda}_1 & \lambda_1 & \bar{\lambda}_1 \\ 1+\lambda_2\bar{\lambda}_2 & \lambda_2 & \bar{\lambda}_2 \\ 1+\lambda_3\bar{\lambda}_3 & \lambda_3 & \bar{\lambda}_3 \end{vmatrix} = 0.$$

**10.** Verify that if

$$\lambda_1\bar{\lambda}_2\left(\cosh\frac{c}{k}+\cosh\frac{a}{k}\right)-\sinh\frac{c}{k}(\lambda_1 e^{-ia}+\bar{\lambda}_2 e^{ia})+\cosh\frac{c}{k}-\cosh\frac{a}{k}=0,$$

the points $\lambda_1$, $\lambda_2$ are collinear with the point whose polar coordinates are $(c, a)$.

**11.** Prove that the formula for a hyperbolic radiation, corresponding to a translation in space through distance $d$, is $\sinh x' = e^d \sinh x$.

**12.** Prove that the transformation $\tanh r \tanh r' = \text{const.}$ changes a straight line into a curve of the second degree.

**13.** Prove that the inverse of the absolute in a circle of radius $c$ is a circle of radius equal to $k \log \cosh c/k$. (This circle is called the *vanishing circle*; cf. vanishing plane in the theory of perspective.)

**14.** Prove that the inverse of a straight line is a circle cutting the vanishing circle orthogonally.

**15.** Prove that the inverse of a horocycle is a circle touching the vanishing circle.

**16.** Prove that any circle which cuts the circle of inversion orthogonally is unaltered by inversion.

**17.** Prove that the inverse of a system of parallel lines is a system of circles all touching at the same point.

**18.** Prove that a horosphere which cuts the vanishing sphere orthogonally is inverted into a plane touching the vanishing sphere; and that a horocycle traced on the horosphere is inverted into a circle lying in this plane and passing through the point of contact. Hence deduce that the geometry on the horosphere is euclidean.

**19.** Show that the equations $x+\mu y=kt(z-1)$ and $x+\mu y=kt(z+1)$, where $t=\tanh a/k$, represent two pencils of branches of equidistant-curves, the first passing through the origin, the second through the point on the axis of $x$ at distance $2a$ from the origin. Prove that the inverses of these systems with respect to a circle with centre $O$ and radius $2a$ are respectively $x+\mu y=kt^3(z+1)$ and $t(x+\mu y)=k(z-1)$.

**20.** Prove that the inverse of the pencil of straight lines $x+\mu y=ktz$, where $t=\tanh a/k$, with respect to a circle with centre $O$ and radius $a$, is the pencil of circles $2p^2(x+\mu y)=kt[(p^4+1)z+(p^4-1)]$, where $p=\tanh \frac{1}{2}a/k$. Show that the common points of this pencil are on the axis of $x$ at distances from the origin equal to $a$ and $b$, where $\tanh \frac{1}{2}b/k=p^3$.

**21.** Prove that inversion with regard to the absolute is represented by $\lambda'\bar{\lambda}=1$. Show that this transformation leaves every straight line unaltered, and changes the circle $ax+by+cz+d=0$ into $ax+by+cz-d=0$, *i.e.* interchanges the two branches of an equidistant-curve.

**22.** Prove that two successive inversions in the two branches of an equidistant-curve of distance $k\sinh^{-1}1$, followed by a reflexion in its axis, are equivalent to an inversion in the absolute.

# CHAPTER IX.

## THE CONIC.

1. A conic is a curve of the second degree, *i.e.* one which is cut by any straight line in two points. Since the equation of a straight line in Weierstrass' coordinates is homogeneous and of the first degree, the equation of the conic will be a homogeneous equation of the second degree. In Cayley's representation a conic will be represented by a conic. This is the chief beauty of Cayley's representation, that the degree of a curve is kept unaltered.

The projective properties of a conic are the same as in ordinary geometry, and it is only in metrical properties that there is any distinction. Since metrical geometry is reduced to projective geometry in relation to the absolute conic, the metrical geometry of a conic in non-euclidean space reduces to the projective geometry of a pair of conics. The metrical properties are those which are not altered by any projective transformation which transforms the absolute into itself. The metrical geometry of a conic therefore reduces to a study of the invariants and co-variants of a pair of conics.

We shall confine ourselves here to an enumeration of the different types of conics, and a few theorems relating to the focal properties of the central conics which bear the closest resemblance to those in ordinary geometry.

## 2. Classification of conics.

In euclidean geometry, leaving out degenerate forms, there are three species of conics, according as they cut the line at infinity in real, coincident or imaginary points. These are the hyperbola, the parabola and the ellipse. Also, as a special case of the ellipse, we have the circle, whose imaginary intersections with the line at infinity are the two circular points.

In non-euclidean geometry conics are classified similarly with reference to their intersections with the absolute.

Two conics cut in four points, and reciprocally they have four common tangents. The points and lines which a conic has in common with the absolute are called the *absolute points* and *tangents.* These elements may be all real, or imaginary or coincident in pairs. When two absolute points are coincident, two absolute tangents are also coincident. When two points are real and two imaginary, the same is true for the tangents. When the points are all real, the tangents may be all real or all imaginary. When the points are all imaginary, the conic must be within the absolute (for we need not notice a conic which is wholly ideal), and the tangents are all imaginary.

Conics are therefore classified as follows :

(1) Absolute points and tangents all real.
*Concave hyperbola,* with two real branches concave towards a point between them.

(2) Absolute points all real, absolute tangents all imaginary.
*Convex hyperbola,* with two real branches, resembling an ordinary hyperbola.

(3) Absolute points and tangents all imaginary.
*Ellipse,* a closed curve.

(4) Absolute points and tangents two real and two imaginary.
*Semi-hyperbola,* with one real branch.

(5) Absolute points and tangents two coincident and two real.
*Concave hyperbolic parabola*, two real branches touching the absolute at the same point.

(6) Absolute points two coincident and two real, absolute tangents two coincident and two imaginary.
*Convex hyperbolic parabola*, one real branch and an ideal branch touching the absolute.

(7) Absolute points and tangents two coincident and two imaginary.
*Elliptic parabola*, resembling an ordinary parabola.

(8) Absolute points and tangents three coincident and one real.
*Osculating parabola*, one real branch osculating the absolute at one end.

(9) Absolute points and tangents, two pairs of each real and coincident.
*Equidistant-curve*.

(10) Absolute points and tangents all imaginary and coincident in pairs.
*Proper circle*.

(11) Absolute points and tangents all coincident.
*Horocycle*.

In elliptic geometry the absolute points and tangents are all imaginary, and we have only ellipses and proper circles.

**3.** The four absolute points form a complete quadrangle. The diagonal points form a triangle $C_1C_2C_3$ which is self-conjugate with regard to the conic and the absolute. Every chord through any of these points is bisected at the point. The points $C_1C_2C_3$ are therefore *centres* of the conic, and their joins are the *axes*.

The four absolute tangents form a complete quadrilateral. Its diagonal triangle is formed by the three axes. In euclidean geometry the foci of a conic are the intersections of the tangents from the circular points. These are the absolute tangents, and we call therefore the three pairs of intersections of the absolute tangents the *foci* of the conic. Similarly the three pairs of joints of the absolute points are called *focal lines*.

The polars of the foci with regard to the conic are called *directrices*. Two pass through each centre and are perpendicular to the opposite axis.

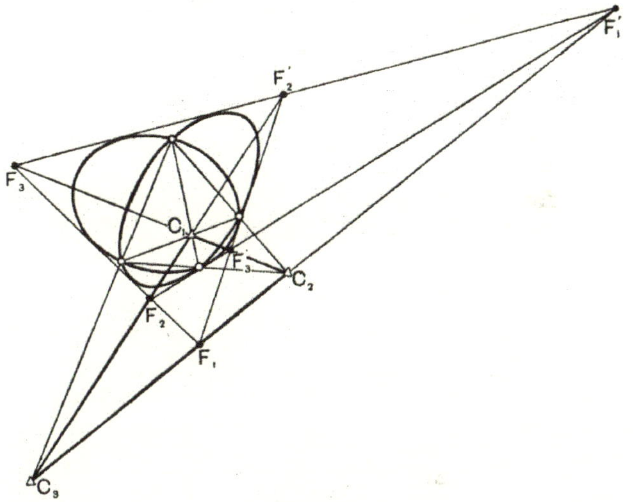


FIG. 126.

The poles of the focal lines with regard to the conic are called *director points*. Two lie on each axis.

In euclidean geometry the focal lines degenerate in two pairs to the line at infinity. The third pair become the asymptotes. Four of the director points coincide with the centre, and the other pair coincide with the points at infinity on the conic. In euclidean geometry the *asymptotes* are the tangents to the conic at the points where it cuts the absolute; but in non-euclidean geometry the lines which most closely resemble the euclidean asymptotes are the tangents to the conic from a centre, and are therefore six in number.

**4.** By taking the triangle formed by the centres as triangle of reference, the equations of the absolute and the conic can be taken in the form

$$x^2+y^2+z^2=0, \quad ax^2+by^2+cz^2=0,$$

or in line-coordinates

$$\xi^2+\eta^2+\zeta^2=0, \quad \frac{\xi^2}{a}+\frac{\eta^2}{b}+\frac{\zeta^2}{c}=0.$$

The coordinates of the common points are given by

$$x^2 : y^2 : z^2 = b-c : c-a : a-b,$$

and the coordinates of the common tangents

$$\xi^2 : \eta^2 : \zeta^2 = a(b-c) : b(c-a) : c(a-b).$$

The focus $F_1$ is the intersection of two absolute tangents

$$\sqrt{a(b-c)}\,x+\sqrt{b(c-a)}\,y-\sqrt{c(a-b)}\,z=0,$$

$$\sqrt{a(b-c)}\,x-\sqrt{b(c-a)}\,y+\sqrt{c(a-b)}\,z=0;$$

therefore its coordinates are

$$0, \quad \sqrt{c(a-b)}, \quad \sqrt{b(c-a)}.$$

$F_1'$ is the intersection of the other pair of absolute tangents, and its coordinates are

$$0, \quad \sqrt{c(a-b)}, \quad -\sqrt{b(c-a)}.$$

If $d$, $d'$ are the distances of a point $P(x, y, z)$ from $F_1$, $F_1'$,

$$\cos d=\frac{y\sqrt{c(a-b)}+z\sqrt{b(c-a)}}{\sqrt{x^2+y^2+z^2}\,\sqrt{a(c-b)}},$$

$$\pm\sin d=i\,\frac{y\sqrt{b(a-b)}+z\sqrt{c(c-a)}}{\sqrt{x^2+y^2+z^2}\,\sqrt{a(c-b)}}.$$

Hence $$\cos(d\pm d')=\frac{\{c(a-b)y^2-b(c-a)z^2\}+\{b(a-b)y^2-c(c-a)z^2\}}{(x^2+y^2+z^2)a(c-b)}$$

$$=\frac{c+b}{c-b}\cdot\frac{(a-b)y^2-(c-a)z^2}{a(x^2+y^2+z^2)}=\frac{c+b}{c-b},$$

i.e. *either the sum or the difference of the distances of any point on a conic from a pair of foci is constant.*

Reciprocally, *either the sum or the difference of the angles which any tangent to a conic makes with a pair of focal lines is constant.*

A tangent makes a triangle with a pair of focal lines. In the case in which the sum of the interior angles is constant the sum of the angles of the triangle is constant, and hence the area is constant. This result may be compared with the property of a hyperbola in euclidean geometry, a tangent to which makes with the asymptotes a triangle of constant area.

5. The conic, the absolute, and a pair of focal lines form three conics passing through the same four points. Any line is cut by these three conics in involution. Let the line cut the conic in $P$, $Q$, the absolute in $X$, $Y$, and the focal lines in $M$, $N$. Then $(XY, PQ, MN)$ is an involution. Let $G$, $G'$ be the middle points of $PQ$, so that

$$(XY, PG)\barwedge(XY, GQ) \quad \text{and} \quad (XY, PG')\barwedge(XY, G'Q).$$

Then $G$, $G'$ are the double points of the involution, and

$$(XY, MG)\barwedge(YX, NG)\barwedge(XY, GN);$$

therefore $G$, $G'$ are also the middle points of $MN$, *i.e. the segments determined by the points of intersection of any line $l$ with a conic and the points in which $l$ cuts a pair of focal lines have the same two middle points.*

Reciprocally, *the tangents from any point $P$ to a conic and the lines joining $P$ to a pair of foci have the same two bisectors.*

If $P$ lies on the conic, *the tangent and normal to the conic at $P$ are the bisectors of $PF$, $PF'$.*

**6.** Take a focus $F$ with coordinates

$$0, \quad \sqrt{c(a-b)}, \quad \sqrt{b(c-a)}.$$

The equation of the corresponding directrix is

$$y\sqrt{b(a-b)}+z\sqrt{c(c-a)}=0.$$

Let $d$ be the distance of any point $P$ on the conic from the directrix and $r$ its distance from the focus; then

$$\sin d=\frac{y\sqrt{b(a-b)}+z\sqrt{c(c-a)}}{\sqrt{x^2+y^2+z^2}\sqrt{(b-c)(a-b-c)}},$$

$$\sin r=\frac{y\sqrt{b(a-b)}+z\sqrt{c(c-a)}}{\sqrt{x^2+y^2+z^2}\sqrt{a(b-c)}}.$$

Therefore $$\frac{\sin r}{\sin d}=\sqrt{\frac{a-b-c}{a}},$$

i.e. *the ratio of the sines of the distances of a point on a conic from a focus and the corresponding directrix is constant.*

Reciprocally, *the ratio of the sine of the angle which a tangent to a conic makes with a focal line to the sine of its distance from the corresponding director point is constant.*

**7.** It is interesting to obtain a geometrical proof of the focal distance property.[1]

Let $\Omega$ be the absolute and $C$ any conic having four real common tangents with $\Omega$. Let two pairs of the common tangents intersect in the pair of foci $F$, $F'$. Let $P$ be any point on $C$. Join $PF$ and $PF'$, cutting $\Omega$ in $X$, $Y$ and $X'$, $Y'$. Then we have to prove that

$$\text{dist. }(PF)\pm\text{dist. }(PF')=\text{const.},$$

or, in terms of cross-ratios,

$$\log(PF, XY)\pm\log(PF', X'Y')=\text{const.},$$

*i.e.* either the product or the quotient of the cross-ratios is constant.

Let $XX'$, $YY'$ cut $FF'$ in $A$ and $B$, $X'Y$ and $XY'$ cut $FF'$ in $A'$ and $B'$.

Then $$(PF, XY)\barwedge_{X'}(F'F, AA')$$

and $$(PF', X'Y')\barwedge_{Y}(FF', A'B)\barwedge(F'F, BA').$$

[1] For part of this proof I am indebted to Dr. W. P. Milne.

Therefore

$$(PF, XY) \div (PF', X'Y') = (F'F, AA') \div (F'F, BA') = (F'F, AB).$$

Similarly $\quad (PF, XY) \,.\, (PF', X'Y') = (FF', A'B').$

We have therefore to prove one of these cross-ratios constant.

Four conics through the points $X, X', Y, Y'$ are $\Omega$; $XY, X'Y'$; $XX', YY'$; $XY', X'Y$. Let $\Omega$ cut $FF'$ in $U, V$. We have then an

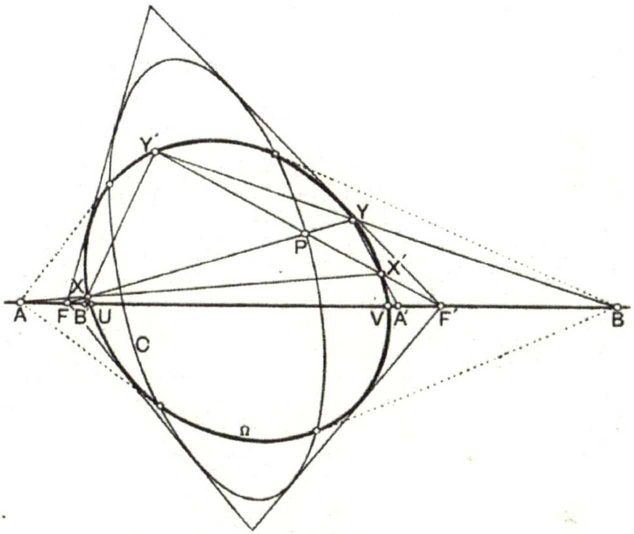


FIG. 127.

involution determined by $(UV, FF')$, and this contains also the pairs $A, B$ and $A', B'$. If therefore $(FF', AB)$ is a given cross-ratio, $A, B$ must be fixed points.

Now, supposing that $A, B$ are fixed points, the point $P$ is constructed thus: $\Omega$ is a fixed conic and $F, F'$ two fixed points. $FF'$ cuts $\Omega$ in fixed points $U, V$, and $A, B$ are a pair of fixed points in the involution determined by $(FF', UV)$.

Through $F$ any line $u$ is drawn cutting $\Omega$ in $X, Y$. $XA$ cuts $\Omega$ again in $X'$, and we get the line $X'F' = u'$ corresponding to $u$. $P$ is the point of intersection of $u$, $u'$. If $F'X'$ cuts $\Omega$ again in $Y'$, then $YY'$ cuts $FF'$ in $B$, the point corresponding to $A$ in the involution $(FF', UV)$.

Since $u$ cuts $\Omega$ in two points, there are two lines $u'$ corresponding to $u$, and similarly there are two lines $u$ corresponding to $u'$. The rays $u$, $u'$ are therefore connected by a (2, 2) correspondence. The locus of $P$ is therefore a curve of the fourth degree. But when $u$ coincides with $FF'$, so also do both the corresponding lines $u'$, and *vice versa*; therefore the locus contains the line $FF'$ twice. It therefore consists of this line doubled and a conic.

Also, if $u$ is a tangent to $\Omega$ the two lines $u'$ coincide, and $P$ is a double point on $u$. Therefore $u$ is a tangent to the locus of $P$. Hence the conic which is the locus of $P$ touches the four tangents drawn from $F$, $F'$ to $\Omega$.

Further, if $P$ is taken on $\Omega$, $X$ and $X'$ coincide with $P$; hence the tangents to $\Omega$ at its points of intersection with $C$ pass through either $A$ or $B$.

$A$, $B$ are therefore the fixed points in which the tangents to $\Omega$ at its intersections with $C$ cut $FF'$, and therefore

$$(PF, XY) \div (PF', X'Y') \text{ is constant.}$$

The foci $F$, $F'$ are real only when the absolute tangents are imaginary.

In the case of the *convex hyperbola* the order of the points $P$, $F$, $X$, $Y$ and $P$, $F'$, $X'$, $Y'$ is the same, and *the difference of the focal distances is constant* (Fig. 128).

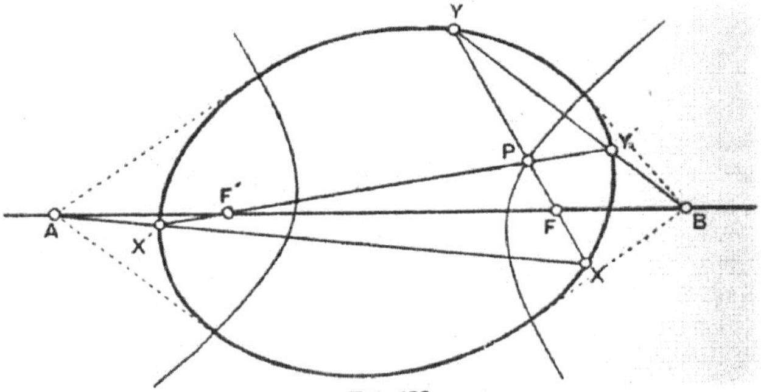


FIG. 128.

In the case of the *ellipse* the points $P, F, X, Y$ and $P, F', Y', X'$ have the same order, and *the sum of the focal distances is constant* (Fig. 129).

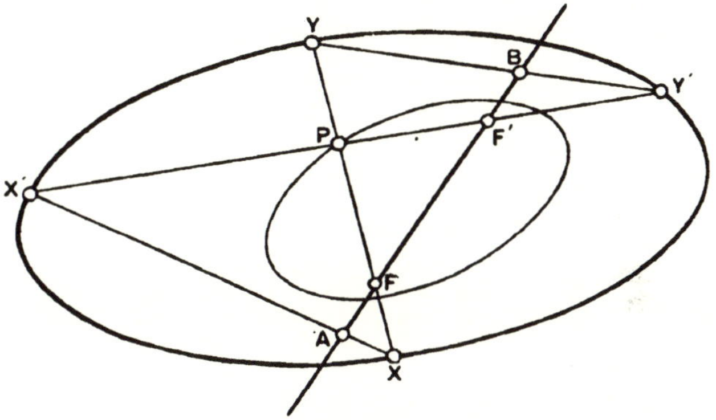


FIG. 129.

## EXAMPLES IX.

**1.** If the equation of the absolute is $x^2+y^2+k^2z^2=0$, prove that the coordinates of the three pairs of foci of the conic

$$x^2/a+y^2/b+k^2z^2/c=0$$

are
$$\left(0,\ \pm k\sqrt{-\frac{\gamma}{a}},\ \sqrt{-\frac{\beta}{a}}\right),\ \left(\pm k\sqrt{-\frac{\gamma}{\beta}},\ 0,\ \sqrt{-\frac{a}{\beta}}\right),$$
$$\left(k\sqrt{-\frac{\beta}{\gamma}},\ \pm k\sqrt{-\frac{a}{\gamma}},\ 0\right),$$

where
$$\alpha=b-c,\ \beta=c-a,\ \gamma=a-b.$$

**2.** In hyperbolic geometry, where the equation of the absolute is $x^2+y^2-k^2z^2=0$, show that the equation $x^2/a+y^2/b-k^2z^2/c=0$ represents (1) an imaginary conic if $a>0$, $b>0$, $c<0$, (2) a real ellipse if $a$, $b$, $c$ are positive and $c$ does not lie between $a$ and $b$, (3) an ideal ellipse if $a<0$, $b>c>0$ or $b<0$, $a>c>0$, (4) a concave hyperbola if $a$, $b$, $c$ are all positive and $c$ lies between $a$ and $b$, (5) a convex hyperbola if $a<0$, $c>b>0$ or $b<0$, $c>a>0$.

3. In elliptic geometry, prove that an ellipse, real or imaginary, has always one pair of real and two pairs of imaginary foci.

4. In hyperbolic geometry, prove that the three pairs of foci are (1) one real and two imaginary for a real or imaginary ellipse or a convex hyperbola, (2) all ideal for an ideal ellipse or a concave hyperbola.

5. $A$, $B$ are fixed points and $APB$ is a right angle; show that the locus of $P$ is an ellipse. If $AB=2a$, prove that the real foci are on $AB$ at a distance from $O$, the middle point of $AB$, such that

$$\tanh x/k = \tanh^2 a/k, \quad \text{or} \quad x = \tfrac{1}{2}k \log \cosh 2a/k.$$

Hence prove the following construction for the foci: Draw $OR$ making the angle $AOR = \Pi(a)$ and cutting the circle on $AB$ as diameter in $R$, $R'$. Then $F$, $F'$ are the feet of the perpendiculars on $AB$ from $R$, $R'$.

6. $A$, $B$ are fixed points and $P$ is a variable point, such that the angle $APB$ is constant; prove that the locus of $P$ is a curve of the fourth degree.

7. $A$, $B$ are fixed points and $P$ is a variable point, such that

$$\cosh \frac{AP}{k} \cosh \frac{BP}{k}$$

is constant; prove that the locus of $P$ is an ellipse.

8. Prove that the locus of a point, such that the ratio of the sines of its distances from two fixed points is constant, is a conic.

9. $A$, $B$ are fixed points and $P$ is a variable point, such that the sum or the difference of the angles $ABP$, $BAP$ is constant; prove that in each case the locus of $P$ is a conic passing through $A$ and $B$.

10. A variable line cuts off on two fixed axes intercepts whose sum or difference is constant; prove that in each case the envelope of the line is a conic touching the axes.

11. Prove that the product of the sines of the distances from a pair of foci to a tangent is constant. State the reciprocal theorem.

12. Prove that the locus of points from which tangents to a central conic are at right angles is a conic meeting the given conic where it meets its directrices. State the reciprocal theorem.

**13.** Prove that the locus of a point which makes with two given points a triangle, whose perimeter is constant, is a conic with the two given points as foci. Show that the locus is also a conic if the excess of the sum of two sides over the third side is constant.

**14.** Prove that the envelope of a line which makes with two given lines a triangle of constant area is a conic. Show that the envelope is also a conic if the excess of the sum of two angles of the triangle over the third is constant.

**15.** Prove that $\dfrac{x^2}{a+\lambda}+\dfrac{y^2}{b+\lambda}+\dfrac{z^2}{c+\lambda}=0$ represents, for all values of $\lambda$, a system of confocal conics.

**16.** Show that in the conformal representation, in which straight lines are represented by circles, a conic is represented by a quartic curve having two nodes at the circular points, *i.e.* a bicircular quartic.

# INDEX

The numbers refer to the pages, except those preceded by **Ex.**, which refer to the examples; *n* signifies footnote.

GLASGOW: PRINTED AT THE UNIVERSITY PRESS BY ROBERT MACLEHOSE AND CO. LTD.

Spotted something that isn't quite right?

Let us know and we'll fix it for future reprints. And as a thank you for your diligence, we'll send you a free Sothis book of your choice.

publisher@sothis.press

This page intentionally left blank

Except of course for that line above. And this one.
And whilst we have your attention, perhaps you'd
like to follow us on Instagram @sothis.press. Go on.

Made in United States
Orlando, FL
04 November 2026

90202368R00173